DIGITAL COMPUTER ARITHMETIC DATAPATH DESIGN USING VERILOG HDL

T0191680

DIGITAL COMPUTER ARITHMETIC DATAPATH DESIGN USING VERILOG HDL

James E. Stine

KLUWER ACADEMIC PUBLISHERS
Boston / Dordrecht / New York / London

Distributors for North, Central and South America:
Kluwer Academic Publishers
101 Philip Drive
Assinippi Park
Norwell, Massachusetts 02061 USA
Telephone (781) 871-6600
Fax (781) 871-6528
E-Mail: <kluwer@wkap.com>

Distributors for all other countries:
Kluwer Academic Publishers Group
Post Office Box 322
3300 AH Dordrecht, THE NETHERLANDS
Telephone 31 78 6576 000
Fax 31 78 6576 254
E-Mail: <services@wkap.nl>

  Electronic Services <http://www.wkap.nl>

Library of Congress Cataloging-in-Publication Data

Digital Computer Arithmetic Datapath Design Using Verilog HDL
James E. Stine
ISBN 1-4020-7710-6

Additional material to this book can be downloaded from http://extras.springer.com

Printed on acid-free paper.

Contents

Preface

The role of arithmetic in datapath design in VLSI design has been increasing in importance over the last several years due to the demand for processors that are smaller, faster, and dissipate less power. Unfortunately, this means that many of these datapaths will be complex both algorithmically and circuit-wise. As the complexity of the chips increases, less importance will be placed on understanding how a particular arithmetic datapath design is implemented and more importance will be given to when a product will be placed on the market. This is because many tools that are available today, are automated to help the digital system designer maximize their efficiently. Unfortunately, this may lead to problems when implementing particular datapaths.

The design of high-performance architectures is becoming more compli-cated because the level of integration that is capable for many of these chips is in the billions. Many engineers rely heavily on software tools to optimize their work, therefore, as designs are getting more complex less understanding is going into a particular implementation because it can be generated automati-cally. Although software tools are a highly valuable asset to designer, the value of these tools does not diminish the importance of understanding datapath ele-ments. Therefore, a digital system designer should be aware of how algorithms can be implemented for datapath elements. Unfortunately, due to the complex-ity of some of these algorithms, it is sometimes difficult to understand how a particular algorithm is implemented without seeing the actual code.

The goal of this text is to present basic implementations for arithmetic data-path designs and the methodology utilized to create them. There are no control modules, however, with proper testbench design it should be easy to verify any of the modules created in this text. As stated in the text, this text is not a book on the actual theory. Theory is presented to illustrate what choices are made and why, however, a good arithmetic textbook is probably required to accompany this text. Utilizing the Verilog code can be paramount along with a textbook on arithmetic and architecture to make ardent strides towards understanding many of the implementations that exist for arithmetic datapath design.

Wherever possible, structural models are implemented to illustrate the design principles. The importance for each design is on the algorithm and not the circuit implementation. Both algorithmic and circuit trade-offs should be adhered to when a design is under consideration. The idea in this text is implement each design at the RTL level so that it may be possibly implemented in many different ways (i.e. standard-cell or custom-cell).

This text has been used partially as lecture notes for a graduate courses in Advanced VLSI system design and High Speed Computer Arithmetic at the Illinois Institute of Technology. Each implementation has been tested with several thousand vectors, however, there may be a possibility that a module might have a small error due to the volume of modules listed in this treatise. Therefore, comments, suggestions, or corrections are highly encouraged.

I am grateful to many of my colleagues who have supported me in this endeavor, asked questions, and encouraged me to continue this effort. In particular, I thank Milos Ercegovac and Mike Schulte for support, advice, and interest. There are many influential works in the area of computer arithmetic that have spanned many years. The interested reader should consult frequent conference on arithmetic such as the IEEE Symposium on Computer Arithmetic, International Conference on Application-Specific Systems, Architectures, and Processors (ASAP), Proceedings of the Asilomar Conference on Signals, Systems, and Computers, and the IEEE International Conference on Computer Design (ICCD) among others.

I would also like to thank my students who helped with debugging at times as well general support: Snehal Ajmera, Jeff Blank, Ivan Castellanos, Jun Chen, Vibhuti Dave, Johannes Grad, Nathan Jachimiec, Fernando Martinez-Vallina, and Bhushan Shinkre. In addition, a special thank you goes to Karen Brenner at Synopsys, Inc. for supporting the book as well.

I am also very thankful to the staff at Kluwer Academic Publishers who have been tremendously helpful during the process of producing this book. I am especially thankful to Alex Greene and Melissa Sullivan for patience, understanding, and overall confidence in this book. I thank them for their support of me and my endeavors.

Last but not least, I wish to thank my wife, Lori, my sons, Justyn, Jordan, Jacob, Caden, and my daughter Rachel who provided me with support in many, many ways including putting up with me while I spent countless hours writing the book.

J. E. Stine

Chapter 1

MOTIVATION

Verilog HDL is a Hardware Description Language (HDL) utilized for the modeling and simulation of digital systems. It is designed to be simple, intuitive, and effective at multiple levels of abstraction in a standard textual format for a variety of different tools [IEE95]. The Verilog HDL was introduced by Phil Moorby in 1984 at the Gateway Design Automation conference. Verilog HDL became an IEEE standard in 1995 as IEEE standard 1364-1995 [IEE95]. After the standardization process was complete and firmly established into the design community, it was enhanced and modified in the subsequent IEEE standard 1364-2001 [IEE01]. Consequently, Verilog HDL provides a fundamental and rigorous mathematical formalistic approach to model digital systems.

The framework and methodology for modeling and simulation is the overall goal of the Verilog HDL language. The main purpose of this language is to describe two important aspects of its objective:

- **Levels of System Specification** - Describes how the digital system behaves and the provides the mechanism that makes it work.

- **System Specification Formalism** - Designers can utilize abstractions to represent their Very Large Scale Integration (VLSI) or digital systems.

1.1 Why Use Verilog HDL?

Digital systems are highly complex. At the most detailed level, VLSI designs may contain billions of elements. The Verilog language provides digital designers with a means of describing a digital system at a wide range of levels of abstraction. In addition, it provides access to computer-aided design tools to aid in the design process at these levels.

The goal in the book is to create computer arithmetic datapath design and to use Verilog to precisely describe the functionality of the digital system. Verilog provides an excellent procedure for modeling circuits aimed at VLSI implementations using place and route programs. However, it also allows engineers to optimize the logical circuits and VLSI layouts to maximize speed and minimize area of the VLSI chip. Therefore, knowing Verilog makes design of VLSI and digital systems more efficient and is a useful tool for all engineers.

Because the Verilog language is easy to use and there are many Verilog compilers publicly as well as commercially available, it is an excellent choice for many VLSI designers. The main goal for this text is to utilize the Verilog HDL as a vehicle towards the understanding of the algorithms in this book. Although there are many algorithms available for the computation of arithmetic in digital systems, a wide variety of designs are attempted to give a broad comprehensive of many of main ideas in computer arithmetic datapath design. Several major areas of digital arithmetic, such as the residue number system, modular arithmetic, square-root and inverse square root, low-power arithmetic, are not implemented in this text. On the other hand, these algorithms are just as important in the area of digital arithmetic. The hope is to eventually have these implementations in future editions of this text where this text focuses on the basic implementations of addition, subtraction, multiplication, and division.

1.2   What this book is not : Main Objective

This book attempts to put together implementations and theory utilizing Verilog at the RTL or structural level. Although the book attempts to always rely on the use of theory and its practical implementation, its main goal is **not** meant to be a book on computer arithmetic theory. On the other hand, this book attempts to serve as a companion to other text books that have done a fantastic job at describing the theory behind the fascinating area of computer arithmetic [EL03], [Kor93], [Mul97]. Moreover, the compendium of articles in [Swa90a], [Swa90b] assembles some of the seminal papers in this area. This reference is particularly illuminating since many of the articles give insight into current and future enhancements as well as being edited for content.

There are a large number of tools that exist in the Electronic Design Automation (EDA) field that attempt to automate many of the designs presented in this book. However, it would be inconceivable for an EDA vendor to be able to generate a tool that can generate every possible arrangement and/or enhancement for digital computer arithmetic. Consequently, this book attempts to illustrate the basic concepts through Verilog implementations. With this knowledge, engineers can potentially exploit these EDA tools at their fullest extent and, perhaps, even augment output that these tools generate. More importantly, it would not be advantageous nor wise for a VLSI engineer to rely on an EDA tool to generate an algorithm without its understanding.

In addition to being able to understand the implementations described in this text, this book also illustrates arithmetic datapath design at the structural level.

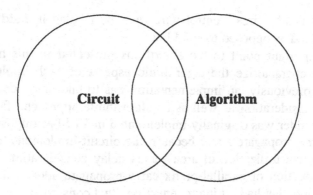

Figure 1.1. Objective for VLSI Engineer.

This allows the designs to be easily visible and allow the reader to relate how the theory associates with the implementation. Moreover, the designs in this book also illustrate Verilog models that are easily importable into schematic-entry VLSI tools. Since VLSI design is about trade-offs between algorithm and circuit implementations as illustrated in Figure 1.1, it is extremely important for a VLSI designer to know what the basic implementation ideas are so that he/she can make an informed decision about a given design objective. In Figure 1.1, the design goal is to form an implementation that makes the best compromise between algorithmic complexity and circuit constraint for a given cost (i.e. the intersection of the two constraints). Therefore, the ideas presented in this book allow the designer to understand the basic implementation in terms of simple gate structures. With this knowledge, the engineer can then apply specific circuit considerations to allow their design to meet a given performance. Without this knowledge, an engineer might lose sight of what an engineer can and can not do with a given implementation.

1.3 Datapath Design

The designs in the book are mainly aimed at design validation. That is, does the design execute the right algorithm and is the functionality of the design function or all potential scenarios or corners. Consequently, when writing a functional model, many designers separate the code into control and datapath pieces. Datapath operate on multi-bit data flow and have regular communication, however, the wiring in control is often more random. Therefore, different wiring indicates that different tools are typically used to support the different parts. This book only concentrates on the datapath design since its more regular, more involved, and takes more time to design, whereas, control logic is usually straight forward and less involved. In order to differentiate control wires from datapath wires, control wires will be labeled in **bold** and drawn using dash-dotted as opposed to solid lines.

Another important point to the algorithms presented in this book is that it attempts to characterize the algorithmic aspects of each implementation. As explained previously, an implementation has to take into consideration a circuit-level consideration as well as its algorithmic approach. For example, a ripple-carry adder was originally implemented in VLSI as opposed to carry-lookahead carry propagate adder because the circuit-implementations lent itself to an efficient utilization of area versus delay considerations. However, with the introduction of parallel-prefix carry-propagate adders, the ability to implement adders that had bit interconnections that consumed less wire length became possible. Therefore, its important for a designer to be aware of how a given implementation performs algorithmically so that certain circuit benefits can benefit the implementation.

To help in characterizing the designs in this book, each implementation will be evaluated for area and delay. To make sure that each implementation is compared without any bias, a given design will utilize AND, OR, and NOT gates only unless otherwise stated. This way, each algorithmic implementations can be compared without confusing an implementation with a circuit benefit. Once the algorithmic advantages and disadvantages are known, a VLSI designer can make astute judgments about how a circuit implementation will benefit a given algorithm. Utilizing additional gates structures, such as exclusive-or gates, would be considered a circuit-level enhancement. Power, an important part of any design today, is not considered in this treatise. Moreover, the designs in this book that require timing control will employ simple D flip flop-based registers although a majority of designs today employ latches to make use of timing enhancements such as multi-phase clocking and/or time borrowing.

For some designs, area will be calculated based on number of gates comprised of basic AND, OR, or NOT gate. Delay is somewhat more complicated and will be characterized in Δ delays using the values shown in Table 1.1. Although one could argue that the numbers in Table 1.1 are not practical, they give each implementation a cost that can allow a good algorithmic comparison. Conceivably, all implementations could even be compared using the same delay value. Altering these delays numbers so that they are more practical starts addressing circuit-level constraints.

Gate	Delay (Δ)
NOT	1
AND	2
OR	2

Table 1.1.   Area and Gate Baseline Units.

Chapter 2

VERILOG AT THE RTL LEVEL

This chapter gives a quick introduction to the Verilog language utilized throughout this book. The ideas presented in this text are designed to get the reader acquainted with the coding style and methodology utilized. There are many enhancements to Verilog that provide high-level constructs that make coding easier for the designer. However, the ideas presented in this book are meant to present computer arithmetic datapath design with easy to follow designs. Moreover, the designs presented here in hopes that the designs could be either synthesized or built using custom-level VLSI design [GS03]. For more information regarding Verilog, there are a variety of books, videos, and even training programs that present Verilog more in depth as well as in the IEEE 1364-2001 standard [IEE01].

2.1 Abstraction

There are many different aspects to creating a design for VLSI architectures. Breaking a design into as much hierarchy as possible is the best way to organize the design process. However, as an engineer sits down to design a specific data path, it becomes apparent that there may be various levels of detail that can be utilized for a given design. Consequently, a designer may decide to design a carry-progagate adder using the "+" operator or using a custom-built 28 transistor adder cell. This level of thought is called an abstraction.

An abstraction is defined as something left out of a description or definition. For example, one could say "an animal that meows" or a "cat". Both could be abstractions for the same thing. However, one says more about the noun than the other. If a VLSI or HDL designer is going to be relatively sane by the time his/her next project comes around, this idea of abstraction should be utilized. That is, in order to be able to handle the large complexity of the design, the designer must make full use of hierarchy for a given design. In other words, it should be broken into three steps as shown below. Hierarchy also exemplifies design reuse where a block may be designed and verified and then utilized many different places.

- Constrain the design space to simplify the design process
- Utilize hierarchy to distinguish functional blocks
- Strike a balance between algorithmic complexity and performance.

This idea is typically illustrated by the use of the Gajski-Kuhn (GK) Y-chart [GK83] as shown in Figure 2.1. In this figure, the lines on the Y-chart represent three distinct and separate design domains – the behavior, structural, and physical. This illustration is useful for visualizing the process that a VLSI designer goes through in creating his/her design. Within each of these domains, several segmentations can be created thus causing different levels of design abstraction within a domain. If the design is created at a high level of abstraction it is placed further away from the center of the chart. The center of the chart represents the final design objective. Therefore, the dotted line represents the design process where an engineer arbitrarily goes from the Behavior, Structural, to the Physical domains until he/she reaches the design objective. Its important to understand that the point at which an engineer starts and ends in the GK chart is chosen by the designer. How he/she achieves their objective is purely a offspring of their creativity, artistic expression, and talent as an engineer. Two engineers may arrive at the same point, however, their journey may be distinct and different. In reality, this is what drives the engineering community. It is our expressionism as a designer in terms of placing his/her creativity into a design. Every designer is different giving a meaning and purpose for each design.

For the "process" to start, a designer picks a location where he/she will start his her design cycle. Many designers are tempted to pick a level further away from the center of the chart such as a behavior level. Although some designers try to explain the reasoning behind this decision being that it saves them time trying to represent their design within a given time construct, this tendency may lead to designs that are less suitable for synthesis. On the other hand, current and future synthesis packages are becoming more adept towards using higher level abstractions, since CAD tools are utilizing better code density and data structure use. Therefore, these new trends in better synthesis packages are giving designers more freedom within their design.

This idea of invoking better synthesis packages is especially true today where feature sizes are becoming smaller and the idea of a System on a Chip (SoC) is in the foremost thought of corporations trying to make a profit. However, the true reasoning behind this is that Verilog, which is a programming language, is built with many software enhancements. These enhancements al-

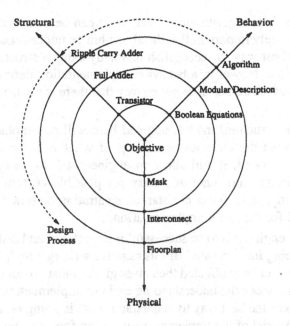

Structural   Behavior

Ripple Carry Adder

Full Adder   Algorithm

Transistor   Modular Description

Boolean Equations

Objective

Mask

Interconnect

Design
Process

Floorplan

Physical

Figure 2.1. GK Y-Chart.

low the designer more freedom in their language use. However, it is important
to realize that Verilog is really meant for digital system design and is not really
a true programming language – it is a modeling language! Consequently, when
a designer attempts to create a design in Verilog, the programmer can code the
design structurally or at the RTL level where an engineer specifies each adder's
organization for a parallel multiplier. Conversely, an engineer may attempt to
describe a system based on a high level of abstraction or at the behavioral level.

As explained previously, the process of going from one domain to another
is termed the *design process*. Behavioral descriptions are transformed to
structural descriptions and eventual to physical descriptions working their way
closer and closer towards the center of the GK chart. After arriving at the
physical domain, an engineer may decide to go back to the behavioral domain
to target another block. However, it is important to additionally realize that a
design may be composed of several abstractions. That is, the multiplier mentioned previously may have some structural definitions, some behavioral definitions, and even some physical definitions. The design process has an analogy
as a *work* factor. The further away from the desired ending point, the harder
the engineer or synthesis program has to work.

In this book, a pure structural form is adopted. This is done for several reasons. The main reason is to construct an easy to understand design for a given
algorithm. If the VLSI engineer is to understand how to make compromises

between what is done algorithmically and what can be achieved with a given circuit, it is extremely important that the algorithm be understood for its implementation. The best way to accomplish this is by using a structural model for an implementation. However, a behavior implementation definitely can have its advantages and the author does not convey that there is no use for a behavior description.

Therefore, the structural models adopted here will accomplish two things. First, it will enable the clear understanding of what is being achieved with a given design. Second, it will allow an engineer an easy way to convert a given Verilog design into schematic-entry for possible custom-level integration. Consequently, an engineer can start concentrating on how that circuit can be implemented for the given implementation.

Overall, if an engineer is to be successful it important that he/she understand the algorithm being implemented. If this requires writing a high-level program to understand how error is affected then an engineer must go out of their way to create a bridge between the understanding and the implementation. Therefore, the author believes the best way to understand what is going on is to attempt a solid structural model of the implementation. Therefore, this book attempts to design each block at the structural level so its easily synthesizable or imported into a schematic for a custom-level design flow.

Moreover, there are various different design methodologies that can be implemented to create the designs in this book as well as other designs elsewhere. Subsequently, using a structural model is not the only solution, however, it definitely is an easy way to understand how a given algorithm can be implemented. The IEEE has attempted to make this more succinct clear in a recent addendum to the standardization so that is better suited towards synthesis [IEE02].

2.2   Naming Methodology

The use of naming conventions and comments are vital to good code structure. Every time a design is started, it is a good idea to keep note of certain features such as comments, radix or base conversion for a number, number of gates, and type of gate with the structure of the design. In this book, the copious use of naming conventions and enumeration is utilized. The naming methodologies listed below are recommended by the author for good program flow, however, they are not required. Verilog is also case-sensitive, therefore, it matters whether you refer to *Cout* or *cout*. It is also a good idea when coding a designs, to use single line comments to comment code. Two types of comments are supported, single line comments starting with // and multiple line comments delimited by /* ... */.

One of the advantages to the Verilog language is that you can specify different radices easily within a given design. Numbers within Verilog can be in binary (*b* or *B*), decimal (*d* or *D*), hexadecimal (*h* or *H*) or octal (*o* or *O*). Numbers are specified by

```
<size>'<radix><number>
```

The size specifies the exact number of bits used by the number. For example, a 4-bit binary number will have 4 as the size specification and a 4 digit hexadecimal will have 16 as the size specification since each hexadecimal digit requires 4 bits. Underscores can also be put anywhere in a number, except the beginning, to improve readability. Several conversions are shown in Figure 2.2.

```
8'b11100011     // 8 bit number in radix 2 (binary)
8'hF2           // 8 bit number in radix 16 (hexadecimal)
8'b0001_1010    // use of underscore to improve readability
```

Figure 2.2.   Binary or Hexadecimal Numbers.

2.2.1   Gate Instances

Verilog comes with a complete set of gates that are useful in constructing your design. Table 2.1 shows some of the various representations of elements that are comprised into the basic Verilog language. Verilog also adds other complex elements for tri-stating, transistors, and pass gates. Since there are a variety of gates that can be implemented, the designs in this book use only basic digital gates. This simplifies the understanding of the design as well as giving the design more of a structural implementation.

All uses of a device are differentiated by instantiating the device. The instantiated value is used by either the simulator or synthesis program to reference each element, therefore, providing a good naming structure for your instance is critical to debugging. Therefore, if a user decides to incorporate two elements of a NAND device, it must be differentiated by the instance or instantiation as seen below:

```
nand nand1 (out1, in1, in2);
nand nand2 (out2, in2, in3, in4);
```

As seen in this example, a gate may also be abstracted for different inputs. For example, the second instantiation is a 3-input NAND gate as opposed to the first instantiation which is a 2-input NAND gate.

In this book, code is usually instantiated as *gateX* where *gate* is a identifier that the user creates identifying the gate. For example, *nand* in the example above is a NAND gate. The value *X* is the specific number of the gate. For

Gate	Format	Description
and	and (output, inputs)	n-input *and* gate
nand	nand (output, inputs)	n-input *nand* gate
or	or (output, inputs)	n-input *or* gate
nor	nor (output, inputs)	n-input *nor* gate
xor	xor (output, inputs)	n-input *xor* gate
xnor	xnor (output, inputs)	n-input *xnor* gate
buf	buf (output, inputs)	n-input buffer gate
not	not (output, input)	n-input inverter gate

Table 2.1. Gate Instances in Verilog.

example, *nand2* is the second NAND gate. This way, its easy to quantify how many gates are utilized for each implementation.

2.2.2   Nets

A *net* represents a connection between two points. Usually, a net is defined as a connection between an input and an output. For example, if Z is connected to A by an inverter, then it is declared as follows:

```
not i1(Z, A);
```

In this example, Z is the output of the gate, whereas, A is the input. Anything given to A will be inverted and outputted to Z.

However, there are times when gates may be utilized within a larger design. For example, the summation of three inputs, A, B, and Cin, sometimes use two exclusive-or gates as opposed to a three input exclusive-or gate. This presents a problems since the output of the first gate has no explicit way of defining an intermediate node. Fortunately, Verilog solves this with the *wire* token. A *wire* acts as an intermediate node for some logic design between the input and output of a block as shown in Figure 2.3. In this example, the variable *temp_signal* is utilized to connect the *xor2* instantiation to the *xor3* instantiation.

2.2.3   Registers

All variables within Verilog have a predefined type to distinguish their data type. In Verilog, there are two types of data types, nets and registers. Net variables, as explained previously, act like a wire in a physical circuit and establish connectivity between elements. On the other hand, registers act like a

```
input   A, B, Cin;
output  Sum1, Sum2;
wire    temp_signal;

xor xor1 (Sum1, A, B, Cin);
xor xor2 (temp_signal, A, B);
xor xor3 (Sum2, temp_signal, Cin);
```

Figure 2.3. Wire Nets.

variable in a high-level language. It stores information while the program is being executed.

The register data type, or *reg*, definition enables values to be stored. A *reg* output can never be the output of a predefined primitive gate as in Table 2.1. Moreover, it can not be an *input* or *inout* port of a module. It is important to remember that this idea of storage is different than how a program stores a value into memory. Register values are assigned by procedural statements within an *always* or *initial* block. Each register value holds its value until an assignment statement changes it. Therefore, this means that the *reg* value may change at a specific time even without the user noticing it which can give digital system engineers some difficulty.

The use of the register data type is extremely useful for the design of storage elements such as registers and latches. Although there are several different variants on how to create a storage element, it is sometimes hard to write a good piece of HDL code for such a device. This is due to the fact that storage elements such as memory and flip-flops are essential feedback elements. This feedback element usually is complicated with a specific circuit that stores a value and another specific circuit that enables reading or writing.

In an attempt to alleviate some of the differences for synthesis, the IEEE has attempted to define this more efficiently in the new 2002 standard [IEE02]. Although the IEEE has made ardent strides towards addressing these issues, it may be awhile before synthesis programs adhere to this standard. Therefore, its always a good idea to consult manuals or white papers for your synthesis program to obtain a good model for such a device. In this book, a simple D-type flip-flop is utilized as shown in Figure 2.4.

2.2.4   Connection Rules

There are several different types of ports utilized within a design. The ports of a Verilog file define its interface to the surrounding environment. A port's

```
module dff (reg_out, reg_in, clk);

    input   clk;                    // clock
    input   reg_in;                 // input

    output  reg_out;                // output

    reg     reg_out;

    always @(posedge clk)
      begin
        reg_out <= reg_in;
      end

endmodule
```

Figure 2.4.   D Flip Flop.

mode can be unidirectional (input, output) or bidirectional (inout). Bidirectional ports must always be of a *net* type and can not be of type *reg*.

2.2.5   Vectors

Another important element within Verilog is the creation of vectors. A vector in Verilog is denoted by square brackets that encloses a contiguous range of bits.   Both the register and net data types can be any number of bits wide if declared as vectors. The Verilog language specifies that for the purpose of calculating the decimal equivalent value of a vector, the leftmost index in the bit range is the most significant bit, whereas, the rightmost bit is the least significant bit [IEE95]. An expression can be also indexed in part or in its entirety. For example, an 8-bit word *cout* can be referenced as cout[7:0].

2.2.6   Memory

Another useful element in this book is the idea of memory. Memory is invariably defined as behavior-level HDL code. This is because memory, either dynamic or static, is composed of many parts including analog devices such as sense amplifiers. Therefore, most VLSI designers code memory at the behavior-level leaving the implementation up to custom-level designs or specialized memory compilers. Since many of the designs in this book utilize memory, memory will be defined at the behavior-level as well.

Memories are simply an array of registers. The example that is utilized in this book is for read-only memory or ROM as shown in Figure 2.5. In this example, the code basically reads in a file called plain text file $rom.data$ which contains the values stored in memory. This is a simple model where the size of the memory is $2^{address} \times data$.

```
input [4:0]     address;    // address

output [7:0]    data        // data

reg [7:0]       memory[0:31];

initial
begin
  $readmemb(''./rom.data'', memory);
end

assign data = memory[address];
```

Figure 2.5. Memories.

2.2.7 Nested Modules

Complex digital systems are designed by systematically partitioning designs into simpler hierarchical functional units. With this use of hierarchy, a design can be managed and executed efficiently. In the Verilog language, each digital system is described by a set of modules. Each of these modules has an interface to other modules to describe how they are connected. The top level design is instantiated into an architecture where each module is invoked hierarchically. Because each module is separated hierarchically, it allows reuse and makes the design simpler to design. This divide and conquer strategy with the use of abstraction makes the design of millions and even billions of devices possible.

Modules can represent pieces of hardware ranging from simple gates to complete systems. Modules can either be specified behaviorally or structurally (or a combination of the two). The keywords *module* and *endmodule* enclose the Verilog description of the device. The text between these two keywords can be in any order, however, its probably best to make sure *declarations* are before *instances* to improve readability. The structure of a module is as follows:

```
module <module name> (<port list>);
  <declarations>
  <instances>
endmodule
```

The <module name> is an identifier that uniquely names the module. The <port list> is a list of input, inout and output ports which are used to connect to other modules. The <declarations> section specifies data objects as registers, inputs, outputs, and wires. The instances are the individual instantiations of primitive gates such as the gates in Table 2.1. In addition, instances may be modules that can be nested within other modules. When a module is referenced by another module, a hierarchical description of the design is invoked.

When calling a module the width of each port must be the same. However, output ports may remain unconnected. This is sometimes useful if some outputs were designated for debugging purposes or if some outputs of a more general module were not required in a particular context. However input ports cannot be omitted for obvious reasons. On the other hand, both *input* and *output* names must be declared in the *port* list.

An important tip to remember regarding instances and port lists is to always define the ports with output first followed by inputs. Since a module port list can be declared with either inputs or outputs first, this can lead to potential design bugs which are difficult to ascertain. The author has seen countless designs plagued by errors because the design may be semantically correct, however, the port lists were declared and instantiated differently. A compiler may not alert you to this problem when its compiled. Therefore, in order to maintain conformity, its advisable to keep outputs first followed by inputs. This methodology is recommended because the primitive gates in Table 2.1 can only be defined output first. This establishes a common methodology and may potentially reduce debugging time from hours to days.

The semantics of the module construct in Verilog are different from subroutines, procedures and functions in other languages. A module is never called! A module is instantiated at the start of the program and stays around for the life of the program. A Verilog module instantiation is used to model a hardware circuit where we assume no one changes the wiring. Each time a module is instantiated, the instantiation is given a name. For example, Figure 2.6 shows a shows a structural implementation of a NAND gate.

2.3   Force Feeding Verilog : the Test Bench

Verilog has been designed to be intuitive and simple to learn. This is why a programmer may see many similarities between Verilog and other popular high-level languages like Pascal and C. Verilog can closely model real circuits using built-in primitives and user-defined primitives. It can also incorporate

```
module nand(out, in1, in2);

  input   in1, in2;

  output out;

  nand nand1(out, in1, in2);

endmodule
```

Figure 2.6. Structural Model of a 2-input Nand Gate.

timing information into its simulation for accurate timing and skew budget checking.

Before a design can be synthesized, it must be verified. Verification is an extremely important aspect of design especially due to high visibility cases such as the Intel Pentium Bug that occurred during the latter half of the 20th century [CT95]. Therefore, there should be an easy way to test a digital system easily in Verilog. Unfortunately, testing is not for the faint hearted. It involves designing good test cases usually using theories related to Boolean simplification and Design for Testability (DFT).

Fortunately, Verilog has an excellent way to facilitate testing. Although many Verilog compilers come with mechanisms to test Verilog code interactively bit by bit, an engineer should use a simple procedure for completing digital system designs. This involves the following:

1 Identify the algorithm to be utilized.

2 Perform an adequate error analysis to confirm that a particular precision will be maintained.

3 Generate input and expected output test vectors into a "golden" file that give the design a good bit coverage.

4 Utilize an automated testing procedure to compare modeled output versus this "golden" file.

The device that enables the testing to occur automatically is called a testbench [Ber03]. This is illustrated graphically in Figure 2.7.

Although we can represent a function many different ways, either behaviorally or structurally or combinations of both, Verilog code is worthless unless we can test it appropriately. In Verilog, although manual stimulation of an input is possible, the recommended method of testing is through a test bench. A

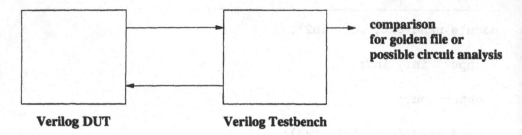

Figure 2.7.   Test Methodology.

test bench is another Verilog module that acts as the stimulus and port watcher for your top-level or part of the Verilog file you wish to simulate. The top-most hierarchy is most-often utilized for testing, since a majority of programs can see below the top level, whereas, if you test at a lower level, the possibility of testing higher-level modules is more difficult.

2.3.1   Test Benches

As stated previously, testing is most often done using test benches or stimulus files. A stimulus file is a file which contains all the vectors you want to test is used in this lab to create the vectors you wish to test. It sort of acts like a tester for your Verilog files by stimulating all the inputs of your **top-level** Verilog file and viewing its outputs.

A test bench is made similar to how the book creates flip flops. A sequential statement is used to indicate the order in which values are to be placed at a specific port. For example, in Figure 2.8 a stimulus file is seen for an instance called **mux21**. Its important to notice that the stimulus file now becomes the top-most hierarchy and instantiates the lower-level device to be tested. In this case, the module is instantiated as *dut1*. Sometimes the Verilog file to be tested is called the Device Under Test (DUT).

Stimulus files are quite easy to create. The stimulus file in Figure 2.8 actually shows two forms of output that will be explained later, however, the important part to remember about stimulus files is that they are synchronized by a *clock*. The *clock* acts the part of the circuit which synchronizes the inputs. Consequently, all inputs are placed using a data type *reg*. Since an output does not have to be synchronized to an input, it just has to be observed. Therefore, all outputs will be of type *wire*. This allows the outputs to appear as soon as the signals are validated. In Figure 2.8 the signal *Clk* acts as the clock to synchronize the output. However, its important to realize that the clock does not have to be utilized within the DUT. It is purely for synchronizing the output to the input.

Since most stimulus files are complicated, it is easiest to copy the input/output ports from your top-most hierarchy and place them in your stimulus file. Afterwards, just rename all input ports of type *reg* and all output ports of type *wire*. It is also important to remember to always keep the *Clk* initialized as a type *reg* so it can be synchronized. That is, always use the same test bench, however, modify the instantiation and definitions to make the appropriate stimulus files

Examining the stimulus file in Figure 2.8 indicates there are three key areas which might need to be modified. The first key area is any area marked with an *initial* tag. An initial statement indicates parts where the code is setup when the program is first executed and then it is never visited again. In the first initial statement in Figure 2.8 it can be seen that the clock period is set at 10 ticks where a 50 % duty ratio is utilized.

In the second initial statement in Figure 2.8, the observed output from the Verilog DUT is recorded in a file called test.out. This is useful to compare the output versus the "golden" file. If the two are equal, then a designer can be satisfied that his or her design works. Careful attention to when the data is written to the output file might also be necessary to make sure the output is in the same format as the "golden" file. In Figure 2.8, the test bench outputs values to its corresponding output file every 5 cycles.

The third area that is crucial to its operation is the third *initial* statement. In this statement, the inputs are asserted. Fortunately, in Verilog, you can assert the inputs in different bases which is one of the powerful elements of the language. In other words, you can give an input a value in any form, octal, hexadecimal, binary, or decimal.

2.4 Other Odds and Ends within Verilog

As stated previously, Verilog comes with many enhancements and language structures to allow a VLSI designer freedom to design. The following are lists of some of the enhancements utilized in this text that may be useful to the reader. For more information regarding some of the other tokens, the reader is advised to consult a comprehensive textbook or even the IEEE standard.

2.4.1 Concatenation

The concatenation operator joins sized nets, registers, bits, and vectors. It is prefaced and appended by the curly braces to signify everything within it shall be concatenated. Verilog will evaluate expressions from left to right. Therefore, the concatenation operator forms a single word from two or more operands from left to right. For example, if the operand A is the bit pattern 1111 and Cin is the bit pattern 1101, then $\{A, B\}$ would produce the bit pattern 1111_1101. The concatenation operator is particularly useful when forming busses.

```
module stimulus;

    reg  Clk;              // Simulate based on clock
    reg  A, B, S;          // Define inputs as reg
    wire Z;                // Define outputs as wires
    integer handle3;       // System values for file output
    integer desc3;

    mux21 dut1 (Z, A, B, S);      // Instantiate your DUT

    initial
      begin
        Clk = 1'b1;
        forever #5 Clk = ~Clk;  // Clock period definition
      end

    initial
      begin
        handle3 = $fopen("test.out");  // Open output file
        #100 $finish;                  // Finish time
      end

    always
      begin
        desc3 = handle3; // Pointer to output file
        #5 \$fdisplay(desc3, "%b %b %b %b", A, B, S, Z);
      end

    initial
      begin
        #10  S = 1'b0;
        #0   A = 1'b0;
        #0   B = 1'b0;
      end

endmodule // stimulus
```

Figure 2.8.   Verilog Test Bench for an Instantiated Device called *mux*21

2.4.2 Replication

Replication can be used as a subset of concatenation to repeat a declaration as many times as specified. It also uses the curly braces and is similar to concatenation except that it takes a certain sequence, which can be a concatenated element, and repeats it a specified number of times. For example, if B is the bit pattern 01, then $\{4\{B\}\}$ would produce four instances or bit patterns of B. In other words, the bit pattern, 01_01_01_01 would be created.

2.4.3 Writing to Standard Output

For certain routine operations Verilog provides system tasks or calls in the form $keyword. The most useful of these is $display. This can be used for displaying strings, expression or values of variables. Since system calls do not produce any logic they are ignored by most synthesis programs. However, a VLSI designer should be aware that these calls may interfere with synthesis.

```
$display("Important Text");
$display($time);
$display(" The sum is %b", sum);
```

The formatting syntax for $display is similar to that of *printf* in the high-level programming language, C. For $display, some of the useful formatting and escape sequences are shown in Table 2.2 and Table 2.3, respectively.

Format	Display
%d or %D	Decimal
%b or %B	Binary
%h or %H	Hexadecimal
%o or %O	Octal
%m or %M	Hierarchical name
%t or %T	Time format

Table 2.2. Output Format Specifications.

2.4.4 Stopping a Simulation

The $finish keyword exits the simulation and passes control to the operating system. The keyword $stop also suspends the simulation and puts Verilog into an interactive mode. The $finish keyword is utilized in the stimulus file to stop the simulation.

Format	Display
\n	newline
\t	tab
\\	print \
\"	print "
%%	print %

Table 2.3.   Output Escape Sequences.

2.5   Timing: For Whom the Bell Tolls

For any given model and simulation, there is an attempt to describing a system. The best way to do this is through the use of ordered levels or hierarchies. These hierarchies are different than the hierarchy available through a given instantiation. This specification allows a model to be **dynamically** defined and also identifies useful ways in which such a system can be defined.

These divisions or specification hierarchies were developed into separate knowledge levels which most languages try to adhere to[ZPK00]. With these ordering levels, a language can develop differences between simulation dynamics. This type of formalism for discrete event systems is typically called a Discrete Event System Specification (DEVS) [ZPK00]. This specification recognizes that the simulation deals with the way the Verilog language behaves over time.

Verilog is a DEVS-style simulator. That is, events are scheduled for discrete times and placed on an ordered-by-time wait queue. The earliest events are at the front of the wait queue and the later events occur later. The simulator removes all the events for the current simulation time and processes them according to the hierarchy level. During the processing, more events may be created and placed in the proper place in the queue for later processing. When all the events of the current time have been processed, the simulator advances time and processes the next events at the front of the queue.

This section discusses two types of explicit timing control Verilog can achieve over when statements are to occur. The first type is delay-based timing in which an expression specifies time between events. The second type is event-based that allows expression to execute based on a specific event such as a clock edge.

2.5.1   Delay-based Timing

This method introduces a delay between when a statement is encountered and when it is executed. It is extremely useful when simulating devices with varying degrees of propagation delay such as in a test bench. This form of timing is simple, however, it is quite often confusing. For example, a simple example is shown in Figure 2.9. When using delay-based timing each statement is addressed based on a continuous time basis. That is, for the value of *#15 b = 1'b1* it can be seen that time it is executed at time 25 and **not** 15.

```
initial begin
  b = 0;           // executed at simulation time 0
  #10 b = 1'b0;    // executed at simulation time 10
  #15 b = 1'b1;    // executed at simulation time 25
  b = 0;           // executed at simulation time 25
end
```

Figure 2.9. Delay-based Timing.

The delay value can also be specified by a constant. A common example is the creation of a clock signal shown previously in the stimulus file. In this example, the first statement initializes the clock, and then the second statement creates the delay of 5 Δ cycles for each value of the clock which goes on indefinitely. A Δ cycle is a common time advance function found within HDL simulators.

```
initial begin
  Clk = 1'b1;
  forever #5 Clk = ~Clk;
end
```

2.5.2   Event-Based Timing

A change in the value of a signal or variable during simulation is referred to as an event. Event-based timing control allows conditional execution based on the another event. Verilog waits on a predefined signal or a user defined variable to change before it executes a specific block.

A sensitivity list is crucial to the design of sequential logic. It places the focus for the language on a particular construct to determine if it changes either by delay, event, or level. However, most of the time the sensitivity list is used conjunctively with an event-based change such as a clock transition. For example, in the D-type flip flop example shown in Figure 2.4. In this example, a user can use *posedge*, *negedge*, or a specific state. However, this must be followed by a 1-bit expression, typically a clock.

2.6   Synopsys DesignWare Intellectual Property (IP)

Synopsys attempted to make datapath simpler for the Verilog user by introducing DesignWare Block IP. This library, formerly called Foundation Library, creates a collection of reusable intellectual property blocks that are tightly coupled with the Synopsys synthesis environment. That is, you can create efficient datapath designs that optimally synthesize when using the Synopsys environment.

The library contains high-performance implementations of intellectual property (IP) for many arithmetic logic functions. The idea of a DesignWare IP attempts to create a nice degree of design automation so that the designer utilizes information hiding where the inner details of the code are hidden, however, the synthesis tools perform various high-level optimizations. Although many of these libraries are encouraged and advisable, it is **not** advisable to implement an arithmetic datapath design without knowing the difference between specific implementations.

DesignWare IP consists of verified, synthesis-enhanced design descriptions. Each of these descriptions represents intellectual property that a designer would like to reuse in their design. In addition, DesignWare allows you to model, compile, and use many different licensing directions within a given file enabling the possibility of creating many different design implementations for a given function. An example of a 16-bit parallel-prefix carry-propagate adder that utilizes a Brent-Kung [BK82b] structure is shown in Figure 2.10. It should be noted that the DesignWare implementation is implemented at a high-level of abstraction. Consequently, a large portion of the information about the actual implementation is somewhat hidden from the user.

This book provides a nice framework for designing arithmetic datapath design using the Verilog Hardware Descriptive Language. Although DesignWare could probably replace many of the algorithms in this book. The author strongly advises against using DesignWare blindly. Without knowing what an algorithm and its implementations effects are can be disastrous. On the other hand, the knowledge obtained through implementing the designs in this book complements the use of DesignWare and can lead to efficient designs.

2.7   Verilog 2001

There have been many numerous advancements to the Verilog language including an update to the IEEE standardization [IEE01]. These advances, specified in the Verilog 2001 standard, have mainly been to promote the wide user-base as well as provide competition with other hardware descriptive languages. Some of the major enhancements with Verilog 2001 that may be helpful to readers of this book are the following listed below [Sut01]. There are 33 major enhancements and many of these enhancement could possibly improve

```
module add (A, B, C);

  parameter width = 16;

  input  [width-1:0]  A,
  input  [width-1:0]  B;
  output [width-1:0] C;
  reg    [width-1:0]    C;

  always @(A or B)
    begin : b0
      /* synopsys resource r0 :
       ops = "a1",
       map_to_module = "DW01_add",
       implementation = "bk";
       */
      C = A + B;
      // synopsys label a1
    end

endmodule // add
```

Figure 2.10. DesignWare Example.

the code in this text. The code in this text is coded to be as compatible with the 1995 standard as well as the 2001 standard.

- the use of a $random statement to implement random generation of inputs. This can be extremely useful in a test bench.

- The wire data type is the default data type as in the original standard. However, with the new standard the default data type can be changed or removed. However, in this text the following convention will be utilized. **In this book, we will make the code easier to view by not showing wires. Some compilers will complain about this and produce an error. Therefore, for most code, it is always good to declare a wire when the bit size is greater than 1. Therefore, if there is a variable in a Verilog example shown in this text and it is not declared, it is a wire**

- Verilog now has the ability to handle generic generation with the addition of a *for* keyword. This option is designed to allow the ability to generate multiple instances of a design.

2.8   Summary

The details of the Verilog hardware descriptive language are presented in this chapter that utilized throughout this book. As explained previously, it is not meant to be a comprehensive listing of the language. On the other hand, it provides a brief introduction on how to detail a digital system within the Verilog HDL at the RTL level. There are many details within the language that may be useful for use in this book, however, they were left off to make details easier to understand. The reader is encouraged to read the IEEE standard as well as consult other texts regarding the language.

Chapter 3

ADDITION

VLSI adders are critically important in digital designs since they are utilized in ALUs, memory addressing, cryptography, and floating-point units. Since adders are often responsible for setting the minimum clock cycle time in a processor, they can be critical to any improvements seen at the VLSI level.

A computer arithmetic system is a system that performs an operation on numbers. Most computer arithmetic systems represent numbers using strings of binary digits. For example, an n-bit unsigned binary integer can be converted as follows:

$$A = \sum_{i=0}^{n-1} a_i \cdot 2^n$$

In fixed point number systems, the position of the binary or radix point is constant. The two most common fixed point representations are integer and fractional. Integer representations are commonly used on general purpose computers, whereas, digital signal processors (DSPs) typically use both integer and fractional representations [EB99]. The most common fixed-point representation is two's complement fractional representation which is utilized throughout this book. In this representation, the most significant bit is referred to as the *sign* bit. If the sign bit is one, the number is negative, otherwise, it is positive. An, n-bit fractional numbers has the following form where a_{n-1} is the sign bit:

$$A = a_{n-1}.a_{n-2} \ldots a_1 a_0$$

The value of an n-bit two's complement binary fraction is the following:

$$A = -a_{n-1} + \sum_{i=0}^{n-2} a_i \cdot 2^{i-n+1}$$

3.1   Half Adders

The most fundamental building block in arithmetic systems is the half adder (HA). A HA takes two bits a_k and b_k and produces a sum bit s_k and a carry bit c_{k+1}. The logic equations for a HA are:

$$
\begin{aligned}
s_k &= \overline{a_k} \cdot b_k + a_k \cdot \overline{b_k} \\
&= a_k \oplus b_k \\
c_{k+1} &= a_k \cdot b_k
\end{aligned}
$$

Using the methodology explained previously in Table 1.1, only *AND*, *OR*, and *NOT* gates are used. This implementation is shown in Figure 3.1. It is important in addressing the relevant benefits for this implementation that both its area and delay affects. Consequently, this HA requires 4 logic gates where each logic element is considered to consume "1" gate element. In addition, this HA has the following critical paths. A critical path is any path between any of its input and outputs where $\triangle$ represents a single gate delay. Its important to note all the critical paths so that a designer can ascertain the worst-case path for an implementation.

$$
\begin{aligned}
a_k, b_k &\rightarrow s_k = 5\,\triangle \\
a_k, b_k &\rightarrow c_{k+1} = 2\,\triangle
\end{aligned}
$$

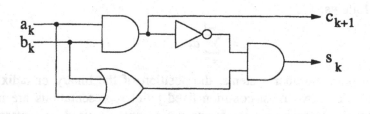

Figure 3.1.   Half Adder (HA) Implementation.

Utilizing the equations formalized above, the Verilog code for the HA is shown in Figure 3.2. As explained previously, variables that are not present in the declaration section are defaulted to *wire* according to the Verilog 2001 standard. It is also worth noting that the instantiation naming convention that is utilized in Figure 3.2 enables a user to easily count the number of gates. That is, 2 *AND* gates, 1 *OR* gate, and 1 inverter.

3.2   Full Adders

A second building block in arithmetic systems is the full adder (FA). A FA takes three bits a_k, b_k, and c_k and produces two outputs a sum bit s_k and a

```
module ha (Cout, Sum, A, B);

    input  A, B;

    output Cout, Sum;

    and a1 (Cout, A, B);
    not i1 (Cbar, Cout);
    or  o1 (p, A, B);
    and a2 (Sum, Cbar, p);

endmodule // ha
```

Figure 3.2. HA Verilog Code.

carry bit c_{k+1}. Sometimes, because a FA counts the number of ones that are available at its input it sometimes is called a $(3, 2)$ counter. The logic equations for a FA are:

$$
\begin{aligned}
s_k &= \overline{a_k} \cdot \overline{b_k} \cdot c_k + \overline{a_k} \cdot b_k \cdot \overline{c_k} + a_k \cdot \overline{b_k} \cdot \overline{c_k} + a_k \cdot b_k \cdot c_k \\
&= a_k \oplus b_k \oplus c_k \\
c_{k+1} &= \overline{a_k} \cdot b_k \cdot c_k + a_k \cdot \overline{b_k} \cdot c_k + a_k \cdot b_k \cdot \overline{c_k} + a_k \cdot b_k \cdot c_k \\
&= a_k \cdot b_k + a_k \cdot c_k + b_k \cdot c_k
\end{aligned}
$$

In this chapter we are considering the implementation of carry-propagate adders (CPA). A CPA produces the result in conventional fixed-radix number system. Another type of adder called a redundant adder is the result is utilized in a redundant number system. This type of adder is discussed later in this text. However, the goal for the adders discussed is to reduce the delay in obtaining carries within a CPA. This idea of studying the carries within a CPA is critical to limiting the delay in the carry generation [Win65], [Win68].

Therefore, an alternate expression for the FA is to utilize the equations based on the relationship between carry-in and carry-out. Consequently, a full adder generates a carry if both a_k and b_k are one. This is called a *generate* and is expressed as follows:

$$g_k = a_k \cdot b_k$$

On the other hand, a full adder propagates a carry if either a_k or b_k is one. This is called a *propagate* and is expressed as follows:

$$p_k = a_k + b_k$$

Therefore, the carry-out equation for the FA can be expressed in terms of the carry-in for a given column.

$$c_{k+1} = g_k + p_k \cdot c_k$$

A FA can be constructed from two HAs and one OR gate, as shown in Figure 3.3. The dotted lines in Figure 3.3 denote the half adder abstraction. This FA requires 9 logic gates and has the following critical paths:

$$a_k, b_k \rightarrow s_k \quad = \quad 10\,\triangle$$
$$a_k, b_k \rightarrow c_{k+1} \quad = \quad 9\,\triangle$$
$$c_k \rightarrow s_k \quad = \quad 5\,\triangle$$
$$c_k \rightarrow c_{k+1} \quad = \quad 4\triangle$$

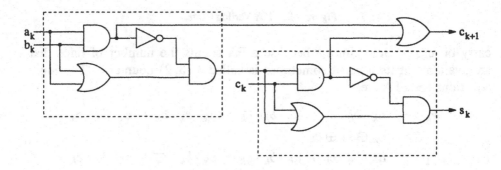

Figure 3.3. Full Adder (FA) Implementation.

Utilizing the equations formalized above, the Verilog code for the FA is shown in Figure 3.4. As expected, the use of hierarchy is utilized and shown by calling the half adder Verilog code shown in Figure 3.2. The use of hierarchy enables the half adder code to be designed once and utilized as needed establishing a strategy of reuse.

3.3   Ripple Carry Adders

Ripple carry adders (RCA) provide one of the simplest types of carry-propagate adder designs. An n-bit RCA is formed by concatenating n FAs. The carry out from the kth FA is used as the carry in of the $(k + 1)^{th}$ FA, as shown in Figure 3.5. The main advantage to this implementation is that it is efficient and easy to construct. Unfortunately, since the connections for the carry-out depend on one another, certain circuit implementations consume a significant amount of delay. On the other hand, because the implementation is simple, certain circuit implementations may be efficiently implemented as a RCA.

```
module fa (Cout, Sum, A, B, Cin);

    input  A, B, Cin;

    output Sum, Cout;

    ha ha1 (g1, temp1, A, B);
    ha ha2 (g2, Sum, temp1, Cin);
    or o1 (Cout, g1, g2);

endmodule // fa
```

Figure 3.4. FA Verilog Code.

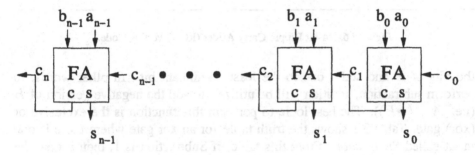

Figure 3.5. Generalized Ripple-Carry Adder (RCA) Implementation.

Since an n-bit RCA requires n FAs and each FA has 9 gates, the total number of gates for the n-bit RCA is $9 \cdot n$. The worst case delays for a ripple carry adder utilizing the critical paths for full adder is generalized according to the number of gates as:

$$a_0, b_0 \rightarrow s_{n-1} = (9 + (n-2) \cdot 4 + 5) \, \Delta$$
$$a_0, b_0 \rightarrow s_{n-1} = (4 \cdot n + 6) \, \Delta$$
$$a_0, b_0 \rightarrow c_n = (4 \cdot n + 5) \, \Delta$$

Utilizing the methodology formalized above, the Verilog code for the 4-bit RCA is shown in Figure 3.4.

3.4 Ripple Carry Adder/Subtractor

Subtraction is just as important as addition to any digital system design. To perform subtraction, it is necessary to take the one's complement each of

```
module rca4 (Sum, Cout, A, B, Cin);

    input [3:0]  A, B;
    input        Cin;

    output [3:0] Sum;
    output  Cout;

    fa fa1 (c0, Sum[0], c0, A[0], B[0], Cin);
    fa fa2 (c1, Sum[1], c1, A[1], B[1], c0);
    fa fa3 (c2, Sum[2], c2, A[2], B[2], c1);
    fa fa4 (c3, Sum[3], c3, A[3], B[3], c2);

endmodule // rca4
```

Figure 3.6.   4-bit Ripple-Carry Adder (RCA) Verilog Code.

the bits of B and add a one to the least significant bit. In other words, to perform subtraction, addition will be utilized to add the negative version of B (i.e. $A + (-B)$). The best logic to perform this function is the exclusive or (xor) gate. Table 3.1 shows the truth table for an xor gate where there is one input called *Subtraction*. From this table, if Subtraction is 1, then it take the one's complement of A, whereas, if Subtraction is 0 it just propagates the value of A.

A	Subtract	Output
0	0	$A = 0$
0	1	$\overline{A} = 1$
1	0	$A = 1$
1	1	$\overline{A} = 0$

Table 3.1.   Exclusive Or Table for Subtraction.

This circuit can then be added to the RCA implementation by placing the XOR gate between the input to B and the full adder. However, in order to support two's complement subtraction, the select signal in Table 3.1 is also input into the C_0 port. Therefore, a row of n XOR gates is inserted to form a ripple-carry adder/subtractor (RCAS), as shown in Figure 3.7. Utilizing the methodology formalized above, the Verilog code for the 4-bit RCAS is shown in Figure 3.8. In this Figure, the port Subtract is inserted removing *Cin* from Figure 3.6.

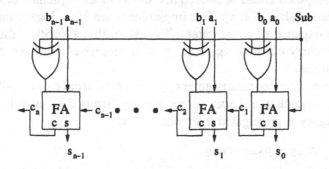

Figure 3.7. Generalized Ripple-Carry Adder/Subtractor (RCAS) Implementation.

```
module rca4s (Sum, Cout, A, B, Subtract);

    input [3:0]  A, B;
    input        Subtract;

    output [3:0] Sum;
    output  Cout;

    xor x1 (w0, B[0], Subtract);
    fa fa1 (c0, Sum[0], A[0], w0, Subtract);
    xor x2 (w1, B[1], Subtract);
    fa fa2 (c1, Sum[1], A[1], w1, c0);
    xor x3 (w2, B[2], Subtract);
    fa fa3 (c2, Sum[2], A[2], w2, c1);
    xor x4 (w3, B[3], Subtract);
    fa fa4 (c3, Sum[3], A[3], w3, c2);

endmodule // rca4s
```

Figure 3.8. 4-bit Ripple-Carry Adder/Subtractor (RCAS) Verilog Code.

3.4.1   Carry Lookahead Adders

The basic idea of the RCA is to let each adder compute a carry and forward it to a subsequent adder. A method to improve the algorithm is to have the carries precomputed ahead of time. This results in an implementation called a carry-lookahead adder (CLA). This implementation has a logarithmic ordered delay at the expense of more gates. To invoke this algorithm the recursive equation for carry-out relating to carry-in is recursively utilized to form all necessary carries.

For example, suppose a carry enters position $k+3$ from carry $k+2$. Utilizing the carry out equation and solving recursively forms the following equation assuming a carry is propagated from position k:

$$
\begin{aligned}
c_{k+3} &= g_{k+2} + p_{k+2} \cdot c_{k+2} \\
&= g_{k+2} + p_{k+2} \cdot (g_{k+1} + p_{k+1} \cdot g_k + p_{k+1} \cdot p_k \cdot c_k) \\
&= g_{k+2} + p_{k+2} \cdot g_{k+1} + p_{k+2} \cdot p_{k+1} \cdot g_k + p_{k+2} \cdot p_{k+1} \cdot p_k \cdot c_k
\end{aligned}
$$

Since the FA implementation discussed earlier does not need to generate the carry for each FA, it can be eliminated. This new implementation shown in Figure 3.9 is called the reduced full adder (RFA). In addition, the FA implementation already has the required logic to produce the generate and propagate. Therefore, the 9-gate FA is reduce to 8-gates consisting of two half-adders with additional outputs for both generate and propagate. In addition, the generate and propagate signals are both ready after $2\triangle$. Since the Verilog code is similar to the FA Verilog code, it is not shown.

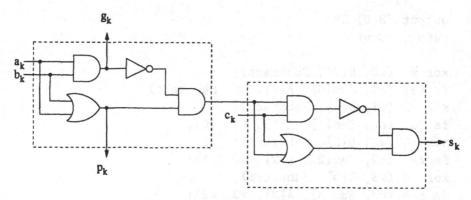

Figure 3.9.   Reduced Full Adder (RFA) Implementation.

The logic used to produce the carries is typically referred to as a carry lookahead generator (CLG). A 4-bit CLG uses 9 gates and the worst-case delay is $4\triangle$. However, some of these gates have higher fan-in requirements. The Verilog code for the 9-gate CLG is shown in Figure 3.10. Notice that the Verilog code could probably be implemented more efficiently if it utilizes a similar approach to the RCA. For example, the code for c_2 could be implemented as follows calling each carry equation recursively:

```
and a2(s2, p[1], cout[1]);
or  o2(cout[2], g[1], s2);
```

The Verilog code for the 9-gate CLG is shown in Figure 3.10. In addition, also note that the use of vectors is utilized to declare the carry-outs as *cout[3:0]* as well as the input propagates and generates (i.e. *g[3:0]* and *p[3:0]*, respectively). This is done for easy debugging, however, single bit declarations are possible.

```
module clg4 (cout, g, p, cin);

    input [3:0]  g, p;
    input        cin;

    output [3:0] cout;

    and a1 (s1, p[0], cin);
    or  o1 (cout[1], g[0], s1);
    and a2 (s2, p[1], g[0]);
    and a3 (s3, p[1], p[0], cin);
    or  o2 (cout[2], g[1], s2, s3);
    and a4 (s4, p[2], g[1]);
    and a5 (s5, p[2], p[1], g[0]);
    and a6 (s6, p[2], p[1], p[0], cin);
    or  o3 (cout[3], g[2], s4, s5, s6);

endmodule // clg4
```

Figure 3.10. Carry Lookahead Generator (CLG) Verilog Code.

Since a 4-bit CLA uses 4 RFAs, and a 4-bit CLG (9 gates), it has a total of $4 \cdot 8 + 9 = 41$ gates. The worst case delay for a 4-bit CLA shown below following the dotted arrow is highlighted in Figure 3.12:

$$a_k, b_k \rightarrow s_3 = 2 + 4 + 5 = 11 \triangle$$
$$a_k, b_k \rightarrow c_4 = 2 + 4 + 4 = 10 \triangle$$

```
module cla4 (Sum, A, B, Cin);

    input [3:0]  A, B;
    input  Cin;

    output [3:0] Sum;

    rfa   r01 (gtemp1[0], ptemp1[0], Sum[0], A[0], B[0],
               Cin);
    rfa   r02 (gtemp1[1], ptemp1[1], Sum[1],A[1], B[1],
               ctemp1[1]);
    rfa   r03 (gtemp1[2], ptemp1[2], Sum[2],A[2], B[2],
               ctemp1[2]);
    rfa   r04 (gtemp1[3], ptemp1[3], Sum[3],A[3], B[3],
               ctemp1[3]);
    clg4 clg1 (ctemp1[3:0], gtemp1[3:0], ptemp1[3:0], Cin);

endmodule // cla4
```

Figure 3.11.   4-bit Carry-Lookahead Adder (CLA) with a CLG Verilog Code.

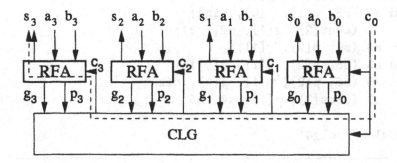

Figure 3.12.   4-bit Carry-Lookahead Adder (CLAA) Implementation that uses a Carry-Lookahead Generator (CLG).

3.4.1.1   Block Carry Lookahead Generators

By examining the equations that result from the recursive equations, each subsequent carry generator increases the fan-in of the logic gates. In VLSI, the increase in fan-in produces gates that require greater amount of driving effort compared to lower fan-in gates [SSH99]. Therefore, carry-lookahead adders beyond 4 bits are not common utilizing CLG logic. To alleviate this problem, the CLG equations are rewritten in terms of blocks (i.e. r). In other words, hierarchy is utilized within the equation to create lookahead logic between each section. For example, the generate and propagate signals are rewritten over 4-blocks as follows:

$$g_{k+3:k} = g_{k+3} + p_{k+3} \cdot g_{k+2} + p_{k+3} \cdot p_{k+2} \cdot g_{k+1} +$$
$$p_{k+3} \cdot p_{k+2} \cdot p_{k+1} \cdot g_k$$
$$p_{k+3:k} = p_{k+3} \cdot p_{k+2} \cdot p_{k+1} \cdot p_k$$

A 4-bit block carry lookahead generator (BCLG) has 14 gates, and a worst-case delay of 4Δ. In summary, the 4-bit block generate and propagate signals can be expressed as

$$c_{k+4} = g_{k+3:k} + p_{k+3:k} \cdot c_k$$

Therefore, a 4-bit CLA with block carry lookahead generator requires $4 \cdot 8 + 14 = 46$ gates.

Utilizing the BCLG logic, a 16-bit CLA can be constructed with 16 RFAs (8 gates each) and 5 BLCG blocks (14 gates each). There is one additional BCLG block to generate the carry-ins for c_4, c_8, and c_{12}. The block diagram is shown in Figure 3.15 . The worst-case critical path is highlighted by the dotted line. There are two important points to pay attention to regarding the CLA logic that utilize BCLG logic. First, the blocks are designed to be equal in this implementation, however, an adder with different block sizes is possible and probably advisable to mitigate wire delay. Second, the delay is shown here from a_0, b_0 to s_7, however, the paths are equal for s_{11}, s_{15}. Therefore, depending on the parasitics associated with this design, the worst-case path could be through either s_7, s_11, or s_15. In summary, the 16-bit CLA requires a total of $16 \cdot 8 + 5 \cdot 14 = 198$ gates. And, the worst-case delay through the 16-bit CLA is the following:

$$a_0, b_0 \rightarrow p_0, g_0 = 2 \, \Delta$$
$$p_0, g_0 \rightarrow g_{3:0} = 4 \, \Delta$$
$$g_{3:0} \rightarrow c_4 = 4 \, \Delta$$
$$c_4 \rightarrow c_7 = 4 \, \Delta$$
$$c_7 \rightarrow s_7 = 5 \, \Delta$$
$$a_0, b_0 \rightarrow c_7 = 19 \, \Delta$$

The Verilog code for the 16-bit CLA that utilizes BCLG logic is shown in Figure 3.14.

```
module bclg4 (cout, gout, pout, g, p, cin);

    input [3:0]  g, p;
    input        cin;

    output [3:0] cout;
    output       gout, pout;

    and a1  (s1, p[0], cin);
    or  o1  (cout[1], g[0], s1);
    and a2  (s2, p[1], g[0]);
    and a3  (s3, p[1], p[0], cin);
    or  o2  (cout[2], g[1], s2, s3);
    and a4  (s4, p[2], g[1]);
    and a5  (s5, p[2], p[1], g[0]);
    and a6  (s6, p[2], p[1], p[0], cin);
    or  o3  (cout[3], g[2], s4, s5, s6);
    and a7  (t1, p[3], g[2]);
    and a8  (t2, p[3], p[2], g[1]);
    and a9  (t3, p[3], p[2], p[1], g[0]);
    or  o4  (gout, g[3], t1, t2, t3);
    and a10 (pout, p[0], p[1], p[2], p[3]);

endmodule // bclg4
```

Figure 3.13. Block Carry-Lookahead Generator (BCLG) Verilog Code.

An n-bit CLA with a maximum fan-in of r, requires

$$\sum_{l=1}^{log_r(n)} \frac{n}{r^l}$$

carry lookahead blocks and n RFAs. An r-bit carry lookahead block requires $\frac{(3+r)\cdot r}{2}$ gates and each RFA requires 8 gates. Thus, the total number of gates for an n-bit CLA is

$$8 \cdot n + \frac{(3+r)\cdot r}{2} \cdot \sum_{l=1}^{log_r(n)} \frac{n}{r^l}$$

In general, an n-bit CLA with a maximum fan-in of r, requires $\lceil \log_r(n) \rceil$ CLA logic levels. An r-bit CLA has $2 + 4 + 5 = 11$ gate delays from the (p, g)

generation, the BCLG logic, and finally for $c_{k+r} \rightarrow s_{k+r}$. From each BCLG, there are 8 additional gate delays per level after the first (i.e. 4 gate delays from the CLG generation and 4 gate delays from the BCLG generation to the next level). Thus, the delay for an n-bit CLAs is

$$11 + 8 \cdot (\lceil \log_r(n) \rceil - 1) = 3 + 8 \cdot \lceil \log_r(n) \rceil$$

```
module cla16 (Sum, G, P, A, B, Cin);

    input [15:0]  A, B;
    input   Cin;

    output [15:0] Sum;
    output        G, P;

    rfa   r01 (gtemp1[0], ptemp1[0], Sum[0], A[0], B[0], Cin);
    rfa   r02 (gtemp1[1], ptemp1[1], Sum[1], A[1], B[1], ctemp1[1]);
    rfa   r03 (gtemp1[2], ptemp1[2], Sum[2], A[2], B[2], ctemp1[2]);
    rfa   r04 (gtemp1[3], ptemp1[3], Sum[3], A[3], B[3], ctemp1[3]);
    bclg4 b1 (ctemp1[3:0], gouta[0], pouta[0], gtemp1[3:0],
                ptemp1[3:0], Cin);
    rfa   r05 (gtemp1[4], ptemp1[4], Sum[4], A[4], B[4], ctemp2[1]);
    rfa   r06 (gtemp1[5], ptemp1[5], Sum[5], A[5], B[5], ctemp1[5]);
    rfa   r07 (gtemp1[6], ptemp1[6], Sum[6], A[6], B[6], ctemp1[6]);
    rfa   r08 (gtemp1[7], ptemp1[7], Sum[7], A[7], B[7], ctemp1[7]);
    bclg4  b2 (ctemp1[7:4], gouta[1], pouta[1], gtemp1[7:4],
                ptemp1[7:4], ctemp2[1]);
    rfa   r09 (gtemp1[8], ptemp1[8], Sum[8], A[8], B[8], ctemp2[2]);
    rfa   r10 (gtemp1[9], ptemp1[9], Sum[9], A[9], B[9], ctemp1[9]);
    rfa   r11 (gtemp1[10], ptemp1[10], Sum[10], A[10], B[10], ctemp1[10]);
    rfa   r12 (gtemp1[11], ptemp1[11], Sum[11], A[11], B[11], ctemp1[11]);
    bclg4 b3 (ctemp1[11:8], gouta[2], pouta[2], gtemp1[11:8],
                ptemp1[11:8], ctemp2[2]);
    rfa   r13 (gtemp1[12], ptemp1[12], Sum[12], A[12], B[12], ctemp2[3]);
    rfa   r14 (gtemp1[13], ptemp1[13], Sum[13], A[13], B[13], ctemp1[13]);
    rfa   r15 (gtemp1[14], ptemp1[14], Sum[14], A[14], B[14], ctemp1[14]);
    rfa   r16 (gtemp1[15], ptemp1[15], Sum[15], A[15], B[15], temp1[15]);
    bclg4  b4 (ctemp1[15:12], gouta[3], pouta[3], gtemp1[15:12],
                ptemp1[15:12], ctemp2[3]);
    bclg4  b5 (ctemp2, G, P, gouta, pouta, Cin);

endmodule // cla16
```

Figure 3.14.   16-bit CLA Verilog Code.

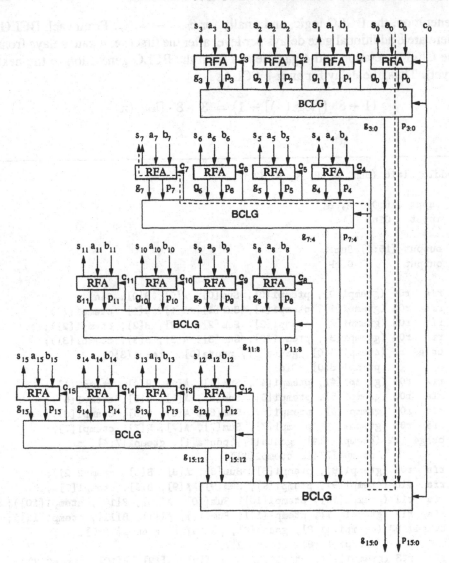

Figure 3.15.   16-bit Carry Lookahead Adder.

3.5   Carry Skip Adders

The carry skip adder (CSKA) is attempt to obtain some of the improvements that were obtained with the CLA. On the other hand, it tries to limit the number of gates it has at the expense of some delay.   In the CSKA, the operands are divided into blocks of r bit blocks. Within each block, a ripple carry adder or smaller CPA is utilized to produce the sum bits and a carry out bit for the block. Again, the CSKA utilizes the carry-out equation expressed in terms of the carry-in for a given column.

$$c_{k+1} = g_k + p_k \cdot c_k$$

From this equation, it can be seen that setting the carry-in signal of a block to zero causes the carry out to serve as a block generate signal. Therefore, an r bit AND gate is also used to form the block propagate signal. The block generate and block propagate signals produce the input carry to the next block. The block diagram for a 16-bit CSKA with 4-bit blocks is shown in Figure 3.15. Once again, the worst-case critical path is highlighted by the dotted line.

In other words, each block tries to detect if a carry is going to bypass the entire smaller CPA block. For example, to obtain the carry into bit position 8, the following equation is utilized:

$$c_8 = g_{7:4} + p_{7:4} \cdot c_4$$

where

$$p_{7:4} = p_7 \cdot p_6 \cdot p_5 \cdot p_4$$

Notice that c_4 can be combined with the group propagate equation with a 5-input AND gate. This adder requires $16 \cdot 9 = 144$ gates to implement the FAs and $2 \cdot 2 = 4$ gates to implement the carry logic, for a total of 148 gates. The delay for this adder is $4 \cdot 4 + 5 = 21\Delta$ to go through the first RCA and $2 \cdot 4 = 8\Delta$ to go through the next two carry-skip blocks, and $4 + 4 + 4 + 5 = 17\ \Delta$ to go through the last RCA block, for a total delay of $46\ \Delta$. The last RCA block takes 17 gate delays instead of $4 \cdot 4 + 6 = 22$ gate delays. This is because the RCA logic has already computed through the first HA block in Figure 3.3 (i.e. it only has to travel from $c_k \rightarrow s_k$). Consequently, the second HA block is waiting for the carry-in. The Verilog code for the 16-bit CSKA is shown in Figure 3.16. The carry skip logic is shown within the $cska16$ module, however, this could easily be incorporated into the $rca4p$ module or a separate module. In addition, the module $rca4p$ code is not shown since it is exactly the same as the $rca4$ module in Figure 3.6 except that each full adder has a respective propagate output signal. However, as its apparent from Figure 3.9, this signal is already produced. Therefore, it just needs to be declared as an *output*.

In general, an n-bit CSKA uses n FAs, each of which requires 9 gates. It also uses $\lceil n/r \rceil - 2$ sets of carry skip logic, each of which requires 2 gates. Thus, the total number of gates used by an n-bit CSKA is:

$$9 \cdot n + 2 \cdot \left(\lceil \frac{n}{r} \rceil - 2 \right)$$

The worst-case delay of an n-bit CSKA uses $(4 \cdot r + 5)\Delta$ for the first block before the carry out is ready. The next $(\lceil n/r \rceil - 2)$ blocks have a delay of 2Δ

```
module cska16 (Sum, Cout, A, B, Cin);

    input [15:0]  A, B;
    input         Cin;

    output [15:0] Sum;
    output        Cout;

    rca4  cpa1 (Sum[3:0],  c4,   A[3:0],   B[3:0],   Cin);
    rca4p cpa2 (Sum[7:4],  w1, p1, A[7:4],  B[7:4],  c4);
    rca4p cpa3 (Sum[11:8], w3, p2, A[11:8], B[11:8], c7);
    rca4  cpa4 (Sum[15:12], Cout, A[15:12], B[15:12], c12);
    and a1 (w2, p1[0], p1[1], p1[2], p1[3], c4);
    or o1 (c7, w1, w2);
    and a2 (w4, p2[0], p2[1], p2[2], p2[3], c7);
    or o2 (c12, w3, w4);

endmodule // cska16
```

Figure 3.16.   16-bit CSKA Verilog Code.

for the carry to skip. The last block has a delay of $4 \cdot r + 1$ from the carry in to the most significant sum bit. Thus, the total delay for s_{n-1} is:

$$4 \cdot r + 5 + 2 \cdot (\lceil \frac{n}{r} \rceil - 2) + 4 \cdot r + 1 = 8 \cdot r + 6 + 2 \cdot \lceil \frac{n}{r} \rceil$$

3.5.1   Optimizing the Block Size to Reduce Delay

The optimum block size is determined by taking the derivative of the delay with respect to r, setting it to zero, and solving for r.

$$8 - \frac{2n}{r^2} = 0$$

$$r = \sqrt{n/4}$$

Plugging this into the delay equation gives

$$8 \cdot \sqrt{n/4} + 6 + 2 \cdot \frac{n}{\sqrt{n/4}} = 4\sqrt{4 \cdot n} + 6$$

For example, if $n = 16$, then the delay is minimized by selecting $r = 2$, which gives a worst case delay of $4\sqrt{4 \cdot 16} + 6 = 38\Delta$. The delay of the carry skip adder can be reduced even further by varying the block size. A good strategy is to use smaller blocks on the two ends and larger blocks in the middle. The design of a 16-bit CSKA with block size of $(1, 2, 3, 4, 3, 2, 1)$ requires 154 gates and has a worst case delay of $34\triangle$. Speed can also be improved by using faster block CPA adders as previously suggested, such as a CLA. In addition, multiple levels of skip logic have also been introduced which also can limit the delay at the expense of more gates. Dynamic programming can also be utilized optimize the block size for single and multiple-levels of skip logic [CSTO92]

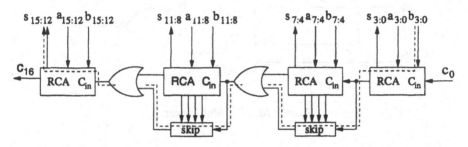

Figure 3.17. 16-bit Carry Skip Adder (r = 4).

3.6 Carry Select Adders

Another popular adder is the carry-select adder (CSEA). The CSEA divides the operands to be added into r bit blocks similar to the CSKA. For each block, except the first, two r-bit ripple carry adders operate in parallel to form 2 sets of sum bits and carry out signals. As in the CSKA, each ripple-carry adder can be replaced by a faster CPA.

Each RCA has two sets of **hard-coded** carry-in signals. One RCA has a carry in of 0, whereas, the other has a carry in of 1. Following the same methodology as the CSKA, the carry in of 0 provides a block generate signal, and carry in of 1 provides a block propagate signal. These two signals are used to generate a carry out signal for the subsequent block. The carry out from the previous block controls a multiplexor that selects the appropriate set of sum bits.

A 2-1 multiplexor, with inputs a and b, select bit s, and output z, can be implemented as

$$z = a \cdot \overline{s} + b \cdot s$$

The 2-1 multiplexor implementation is shown in Figure 3.18 with the worst-case delay highlighted by the dotted line. An r-bit multiplexor can be built using 2-1 multiplexors. An r-bit multiplexor requires $4 \cdot r$ gates and has a worst-case delay of is $5\triangle$. The inversion of s can be accomplished with one gate instead of an inverter for each multiplexor, however, this may cause sizing problems due to drive strength requirements, therefore, for simplicity each 2-1 multiplexor has its own inverter. A 2-bit multiplexor implementation is shown in Figure 3.19. Utilizing the equations formalized above, the Verilog code for the 2-1 multiplexor and the 2-bit multiplexor is shown in Figure 3.20 and Figure 3.21, respectively.

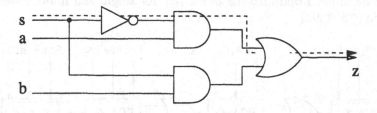

Figure 3.18.   2-1 Multiplexor.

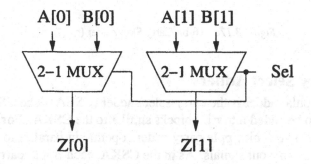

Figure 3.19.   A 2-bit Multiplexor.

A 16-bit CSEA adder with 4-bit blocks is shown in Figure 3.22. Similar to the CSKA, the carry into bit position 8 is obtained by the following equation:

$$c_8 = g_{7:4} + p_{7:4} \cdot c_4$$

where $g_{7:4}$, and $p_{7:4}$ come from the ripple carry adders. The delay for this adder is $4 \cdot 4 + 5 = 21\triangle$ to go through the first ripple carry adder, $2 \cdot 4 = 8\triangle$ to go through the next two blocks, and $5\triangle$ to go through the multiplexor. The total delay is $21 + 8 + 5 = 34\triangle$. The adder requires $4 \cdot 9 + 12 \cdot 9 \cdot 2 = 252$ gates for full adders, $12 \cdot 4 = 48$ gates for the multiplexors, and $2 \times 3 = 6$ gates for the carry logic. The total gate count is $252 + 48 + 6 = 306$ gates.

An n-bit CSEA with r bit blocks uses $2 \cdot n - r$ FAs, each of which requires 9 gates. It uses $\lceil n/r \rceil - 1$ sets of carry logic blocks, each of which requires

```
module mux21 (Z, A, B, S);

   input  A, B, S;

   output Z;

   not i1 (Sbar, S);
   and a1 (w1, A, Sbar);
   and a2 (w2, B, S);
   or o1 (Z, w1, w2);

endmodule // mux21
```

Figure 3.20. 2-1 Multiplexor Verilog Code.

```
module mux21x2 (Z, A, B, S);

   input [1:0]  A, B;
   input        S;

   output [1:0] Z;

   mux21 mux1 (Z[0], A[0], B[0], S);
   mux21 mux2 (Z[1], A[1], B[1], S);

endmodule // mux21x2
```

Figure 3.21. 2-bit Multiplexor Verilog Code.

2 gates for the group propagate and generate logic. The multiplexor logic requires $4 \cdot (n - r)$ gates. Thus, the total number of gates used by an n-bit CSEA is:

$$9 \cdot (2 \cdot n - r) + 2 \cdot \left(\lceil \frac{n}{r} \rceil - 1 \right) + 4 \cdot (n - r) = 22 \cdot n - 13 \cdot r + 2 \cdot \lceil \frac{n}{r} \rceil - 2$$

Similarly, the worst-case delay can be computed as before. For the first RCA, there is a delay of $(4 \cdot r + 5)\Delta$. The next $(\lceil n/r \rceil - 2)$ blocks have a delay of 4Δ for the carry logic and the last RCA block has a delay of 5 for the multiplexor

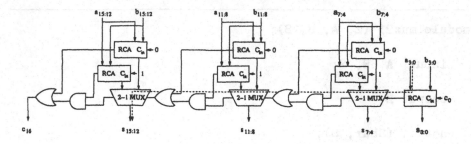

Figure 3.22.   16-bit Carry Select Adder (r = 4).

selection logic. Thus, the total delay for s_{n-1} is:

$$4 \cdot r + 5 + 4 \cdot (\lceil \frac{n}{r} \rceil - 2) + 5 = 4 \cdot r + 4 \cdot \lceil \frac{n}{r} \rceil + 2$$

3.6.1   Optimizing the Block Size to Reduce Delay

The same analysis can be performed to find the optimum block size for a CSEA. The optimum block size is determined by taking the derivative of the delay with respect to r, setting it to zero, and solving for r as done previously.

$$4 - \frac{4 \cdot n}{r^2} = 0$$
$$r = \sqrt{n}$$

Plugging this into the delay equation gives

$$4\sqrt{n} + 4 \cdot \frac{n}{\sqrt{n}} + 2 = 8\sqrt{n} + 2$$

For example, if $n = 16$, then the delay is minimized by selecting $r = 4$, which gives a worst case delay of $8\sqrt{16} + 2 = 34\triangle$ which is actually the same as the implementation above.

Similar to the CSKA, the delay of the CSEA can be reduced even further by varying the block size. The same strategy, by increasing the block size toward the middle of the implementation, as the CSKA is a good methodology for reducing the delay. For example, a 16-bit CSEA with block size of $(2, 2, 3, 4, 5)$ requires 322 gates and has a worst case delay of $30\triangle$. Speed can also be improved by using faster block CPA as previously suggested, such as a CLA.

CSEA logic typically consumes a significant amount of logic. However, as mentioned previously, this type of circuit can be effective when the carry-in arrives later than the input operands. For example, in floating-point units the exponent adder typically has to wait until the carry-in is available later due to

```
module csea16 (Sum, Cout, A, B, Cin);

    input [15:0]  A, B;
    input         Cin;

    output [15:0] Sum;
    output    Cout;

    rca4 rca1 (Sum[3:0], c4, A[3:0], B[3:0], Cin);
    rca4 rca2 (Sum0_0, g4, A[7:4], B[7:4], 1'b0);
    rca4 rca3 (Sum0_1, p4, A[7:4], B[7:4], 1'b1);
    rca4 rca4 (Sum1_0, g8, A[11:8], B[11:8], 1'b0);
    rca4 rca5 (Sum1_1, p8, A[11:8], B[11:8], 1'b1);
    rca4 rca6 (Sum2_0, g12, A[15:12], B[15:12], 1'b0);
    rca4 rca7 (Sum2_1, p12, A[15:12], B[15:12], 1'b1);
    mux21x4 mux1 (Sum[7:4], Sum0_0, Sum0_1, c4);
    and a1 (w1, c4, p4);
    or o1 (c8, w1, g4);
    mux21x4 mux2 (Sum[11:8], Sum1_0, Sum1_1, c8);
    and a2 (w2, c8, p8);
    or o2 (c12, w2, g8);
    mux21x4 mux3 (Sum[15:12], Sum2_0, Sum2_1, c12);
    and a3 (w3, c12, p12);
    or o3 (Cout, w3, g12);

endmodule // csea16
```

Figure 3.23. 16-bit CSEA Verilog Code.

post-normalization. Since the multiplexor only has to be traversed when the carry-in arrives, the CSEA is an efficient adder for these scenarios.

3.7 Prefix Addition

One method of improving carry-propagate adders for computing in logarithmic time is to express it as a prefix computation [BK82a], [HC87], [KS73], [LF80]. Using prefix computations are particularly attractive because it leads to an efficient implementation. In addition, the intermediate structures allow trade-offs between the amount of internal wiring and the fanout of intermediate nodes thereby resulting in a more attractive combination of speed, area and power [Kno01].

Binary carry-propagate adders can be efficiently expressed as a prefix computation [LA94]. That is, through the basic operation of $c_{i+1} = (a_i \cdot b_i) + (a_i + b_i) \cdot c_i$. Parallel prefix logic combines n inputs:

$$x_{n-1}, x_{n-2}, \ldots, x_0$$

using an arbitrary associative operator $\circ$ to n outputs so that the output y_i depends only on inputs $x_{j \leq i}$:

$$
\begin{aligned}
y_0 &= x_0 \\
y_1 &= x_1 \circ y_0 = x_1 \circ y_0 \\
&\vdots \\
y_{n-1} &= x_{n-1} \circ y_{n-1} = x_{n-1} \circ x_{n-2} \circ \ldots \circ x_0
\end{aligned}
$$

The key to fast addition is the fast calculation of the carries c_i [Win65]. Using the recursive equations utilized previously:

$$c_{i+1} = g_i + p_i \cdot c_i$$

with the *generate* or g signal being equal to

$$
g_i = \begin{cases}
a_i \cdot b_i, & if\ 1 \leq i < n \\
a_0 \cdot b_0 + a_0 \cdot c_{in} + b_0 \cdot c_{in}, & if\ i = 0
\end{cases}
$$

and the *propagate* or p signal being equal to

$$p_i = a_i + b_i$$

Some adders utilize propagate as $p_i = a \oplus b$ to exploit specific circuit structures [GSss], [GSH03]. Substituting recursively can be generalized by rewriting the equations for the carry into position $k + r$.

$$c_{k+r} = \left(\sum_{i=k}^{k+r-1} g_i \prod_{j=i+1}^{k+r-1} p_j \right) + c_k \prod_{j=k}^{k+r-1} p_j$$

The final sum can be computed from the carry bits as:

$$s_i = p_i \oplus c_i$$

Defining the operation $\circ$ on an ordered bit pair (g, p)

$$(g_i, p_i) \circ (g_j, p_j) = (g_i + p_i \cdot g_j, p_i \cdot p_j)$$

Using the notation of $\circ$, the recurrence relationship can be rewritten as:

$$(c_{i+1}, p_0 \ldots p_i) = (g_i, p_i) \circ \ldots \circ (g_0, p_0)$$

Therefore, the carries c_i can be calculated using a prefix algorithm (i.e. based on a subscript), however, it is important to point that this new operator $\circ$ is associative and not commutative [BK82a]. Prefix addition is carried out in three consecutive steps called the preprocessing stage, parallel-prefix carry computation, and the postprocessing stage. This is shown in Figure 3.24.

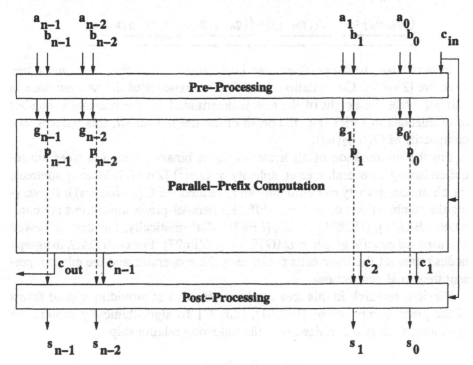

Figure 3.24. Three Stages of a Parallel-Prefix Addition

The parallel-prefix calculation is equivalent to evaluating the new prefix recurrence relations for each bit position, i, for $0 \leq i < n$. However, since the $\circ$ operator is not commutative the order of the operands must not be changed. This makes sense since you can not compute a summand for the upper bits before you can compute the lower bits. However, due to the associativity of the $\circ$ operator, its evaluation does not have be done serially, but can be carried out in any order as shown in the Equation below where each curly brace represents a computation:

$$(g_n, p_n) \circ \underbrace{((g_{n-1}, p_{n-1}) \circ \underbrace{((g_{n-2}, p_{n-2}) \circ g_{n-3}, p_{n-3}))}}$$

$$\underbrace{((g_n, p_n) \circ (g_{n-1}, p_{n-1}))} \circ \underbrace{((g_{n-2}, p_{n-2}) \circ g_{n-3}, p_{n-3})}$$

In particular, the $\circ$ operations can be evaluated according to a binary tree structure [Zim97]. Computations on different branches of the tree are done in parallel while the height of the tree is determined by the maximum number of evaluations in series (i.e. the depth of the tree). Overall, this reduces to a complexity of $O(log_2(n))$.

For the computation of all n carries, c_i, n binary evaluation trees are required having an overall area complexity of $O(n^2)$ [Zim97]. Sharing subtrees, the circuit complexity can be significantly reduced to $O(n \cdot log_2(n))$. By varying the combinations of subtrees, different parallel-prefix algorithms are computed [BK82a], [HC87],[KS73], [LF80]. Mathematically, this can be viewed as a *directed acyclic graph* or *DAG* [LA94], [Zim97]. For each *DAG*, the graph nodes represent the logic cells performing the $\circ$ operator and the edges represent the signal connections.

Previous research in this area has been effective at providing closed-forms of the prefix computations [BSL01], [Zim97] To algorithmically capture the equations, vectors are utilized with the following relationship:

$$v_{i,j} = (g_{i,j}, p_{i,j})$$

Each vector pair denotes the generate, propagate signal pair from the cell (i, j) to the subsequent cell $(i, j + 1)$. Each bit pair is computed as long as the row number is between $1 \leq j \leq h$ where h is the height of the graph.

Based on this new vector notation, $v_{i,j}$, two operators are created. The black cell performs the basic $\circ$ operator as:

$$v_{m,j+1} = v_{m,j} \circ v_{n,j} \ni (m > n)$$

whereas, the white cells simply copy the input to their output. Both cells are shown in Figure 3.25. A cell can have more than one output depending on its drive strength, although, fanouts of 1 or 2 are most common. In other words, the CLA equations are utilized with low block sizes to order the interconnect efficiently. The Verilog code for the white and black cells are shown in Figure 3.26 and 3.27, respectively.

Based on the two new cells, parallel-prefix addition can be computed using simple graphical representations. Various algorithms properties are also visible

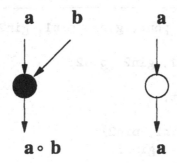

Figure 3.25.   Two Main Cells in Parallel-Prefix Implementations

---

```
module white (gout, pout, gin, pin);

    input gin;
    input pin;

    output gout;
    output pout;

    buf b1 (gout, gin);
    buf b2 (pout, pin);

endmodule // white
```

Figure 3.26.   White Processor Verilog Code.

in each graphs and various rules can be developed based on the associativity of the ○ operator [BSL01], [Zim97]. One of the more popular parallel-prefix algorithms, called the Brent-Kung [BK82b] structure, is shown in Figure 3.28.

The Verilog implementation for Figure 3.28 is fairly intuitive because it follows regular patterns. On the other hand, if non-regular partitions are utilized for the parallel-prefix computation, hybrid schemes can be formulated [Kno99]. Variations on the fanout can also be utilized. In addition, since the ○ groupings can make the interconnect less congested it improves the throughput for sub-micron processes [Kno99], [Kno01]. The Verilog implementation for the Brent-Kung adder is shown in Figure 3.29. Only black processors are shown to save space but they could be easily added. The use of the white processor, although not needed for a prefix implementation, can allow drive strengths to be

```
module black (gout, pout, gin1, pin1, gin2, pin2);

    input  gin1, pin1, gin2, pin2;

    output gout, pout;

    and xo1 (pout, pin1, pin2);
    and an1 (o1, pin1, gin2);
    or  or1 (gout, o1, gin1);

endmodule // black
```

Figure 3.27.   Black Processor Verilog Code.

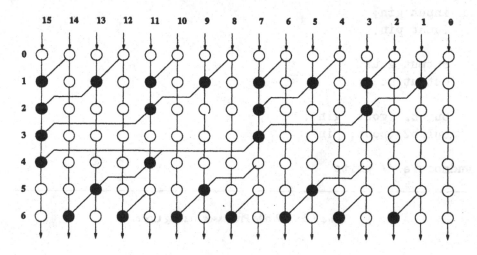

Figure 3.28.   Brent/Kung's Prefix Algorithm

managed more efficiently. The preprocess and postprocess are also not shown to save space.

3.8   Summary

Several adder implementations are discussed in this chapter including ripple-carry addition, carry-lookahead, carry-skip, carry-select, and prefix addition. There are many implementations that utilize hybrid variants of the adders presented in this chapter [GSss], [LS92], [WJM+97]. Figure 3.30 and Figure 3.31 show comparisons of the area and delay for the adders discussed in this chap-

```
module bk16 (Sum, Cout, A, B);

   input [15:0] A, B;

   output [15:0] Sum;
   output   Cout;

   preprocess pre1 (G, P, A, B);
   black   b1 (g0[0], p0[0], G[1], P[1], G[0], P[0]);
   black   b2 (g0[1], p0[1], G[3], P[3], G[2], P[2]);
   black   b3 (g0[2], p0[2], G[5], P[5], G[4], P[4]);
   black   b4 (g0[3], p0[3], G[7], P[7], G[6], P[6]);
   black   b5 (g0[4], p0[4], G[9], P[9], G[8], P[8]);
   black   b6 (g0[5], p0[5], G[11], P[11], G[10], P[10]);
   black   b7 (g0[6], p0[6], G[13], P[13], G[12], P[12]);
   black   b8 (g0[7], p0[7], G[15], P[15], G[14], P[14]);
   black   b9 (g1[0], p1[0], g0[1], p0[1], g0[0], p0[0]);
   black   b10 (g1[1], p1[1], g0[3], p0[3], g0[2], p0[2]);
   black   b11 (g1[2], p1[2], g0[5], p0[5], g0[4], p0[4]);
   black   b12 (g1[3], p1[3], g0[7], p0[7], g0[6], p0[6]);
   black   b13 (g2[0], p2[0], g1[1], p1[1], g1[0], p1[0]);
   black   b14 (g2[1], p2[1], g1[3], p1[3], g1[2], p1[2]);
   black   b15 (g3[0], p3[0], g1[2], p1[2], g2[0], p2[0]);
   black   b16 (g3[1], p3[1], g2[1], p2[1], g2[0], p2[0]);
   black   b17 (g4[0], p4[0], g0[2], p0[2], g1[0], p1[0]);
   black   b18 (g4[1], p4[1], g0[4], p0[4], g2[0], p2[0]);
   black   b19 (g4[2], p4[2], g0[6], p0[6], g3[0], p3[0]);
   black   b20 (g5[0], p5[0], G[2], P[2], g0[0], p0[0]);
   black   b21 (g5[1], p5[1], G[4], P[4], g1[0], p1[0]);
   black   b22 (g5[2], p5[2], G[6], P[6], g4[0], p4[0]);
   black   b23 (g5[3], p5[3], G[8], P[8], g2[0], p2[0]);
   black   b24 (g5[4], p5[4], G[10], P[10], g4[1], p4[1]);
   black   b25 (g5[5], p5[5], G[12], P[12], g3[0], p3[0]);
   black   b26 (g5[6], p5[6], G[14], P[14], g4[2], p4[2]);
   postprocess post2 (Sum, Cout, A, B,
      {g3[1], g5[6], g4[2], g5[5], g3[0], g5[4], g4[1], g5[3],
       g2[0], g5[2], g4[0], g5[1], g1[0], g5[0], g0[0], G[0]});

endmodule // bk16
```

Figure 3.29. 16-bit Brent-Kung Prefix Adder Verilog Code.

ter. As expected, the CLA adder is the fastest algorithmically, however, it consumes the most area. The plots utilize $r = 4$ block sizes. Improvements to CSKA and CSEA designs could be improved by optimizing the block size as discussed in Sections 3.5.1 and 3.6.1. As stated previously, its important to remember that the analysis presented here is algorithmic and does not take into consideration any circuit implementations. Circuit and algorithmic concerns must be addressed together.

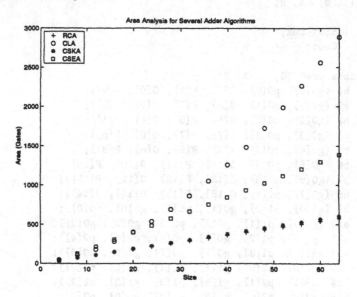

Figure 3.30. Area Plots for Adder Designs.

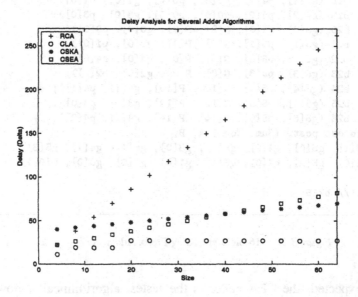

Figure 3.31. Delay Plots for Adder Designs.

Chapter 4

MULTIPLICATION

In this chapter, multiplication datapath designs are explored. As opposed to previous chapters where a design consists of basic gates, the designs in this chapter and subsequent to this chapter utilize designs that are more complex. Consequently, many of the designs that are visited in this chapter and beyond start utilizing more and more of the previous designs made throughout the book. This is a direct use of hierarchy and reuse and is extremely helpful in the design of datapath elements.

Multiplication involves the use of addition in some way to produce a product p from a multiplicand x and multiplier y such that:

$$p = x \cdot y \tag{4.1}$$

High speed multipliers are typically classified into two categories. The first, known as parallel multiplication, involves the use of hardware to multiply a m-bit number by a n-bit number to completely produce a $n + m$ product. Parallel multipliers can also be pipelined to reduce the cycle time and increase the throughput by introducing storage elements within the multiplier. On the other hand, serial or sequential multip'iers compute the product sequentially usually utilizing storage elements so that hardware of the multiplier is reused during an iteration. The implementations presented in this chapter are primarily parallel multipliers since they usually provide the most benefit to a computer architecture at the expense of area. However, many of the designs presented here can be utilized in a sequential fashion as well.

Multiplication usually involve three separate steps as listed below. Although there are implementations that can theoretically be reduced to the generation of the shifted multiples of the multiplicand and multi-operand addition (i.e. the addition of more than two operands), most multipliers utilize the steps below. Although there are various different perspectives on the implementation of multiplication, its basic entity usually is the adder.

1 Partial Product (PP) Generation - utilizes a collection of gates to generate the partial product bits (i.e. $a_i \cdot b_i$).

2 Partial Product (PP) Reduction - utilizes adders (counters) to reduce the partial products to sum and carry vectors.

3 Final Carry-Propagate Addition (CPA) - adds the sum and carry vectors to produce the product.

4.1   Unsigned Binary Multiplication

The multiplication of an n-bit by m-bit unsigned binary integers a and b creates the product p. This multiplication results in m partial products, each of which is n bits. A partial product involves the formation of an individual computation of each bit or $a_i \cdot b_j$. The n partial products are added together to produce a $n + m$-bit product as shown below. This operation on each partial product forms a nice parallelogram typically called a partial product matrix. For example, in Figure 4.1 a 4-bit by 4-bit multiplication matrix is shown. In lieu of each value in the matrix, a dot is sometimes shown for each partial product, multiplicand, multiplier, and product.   This type of diagram is typically called a dot diagram and allows arithmetic designers a better idea of which partial products to add to form the product.

$$
\begin{aligned}
P &= A \cdot B \\
&= \left(\sum_{i=0}^{n-1} a_i \cdot 2^i\right) \cdot \left(\sum_{j=0}^{m-1} b_j \cdot 2^j\right) \\
&= \sum_{i=0}^{n-1}\sum_{j=0}^{m-1} a_i \cdot b_j \cdot 2^{i+j}
\end{aligned}
$$

The overall goal of most high-speed multipliers is to reduce the number of partial products. Consequently, this leads to a reduced amount of hardware necessary to compute the product. Therefore, many designs that are visited in this chapter involve trying to minimize the complexity in one of the three steps listed above.

4.2   Carry-Save Concept

In multiplication, adders are utilized to reduce the execution time. However, from the topics in the previous chapter, the major source of delay in adders is consumed from the carries [Win65]. Therefore, many designers have concentrated on reducing the total time that is involved in summing carries. Since

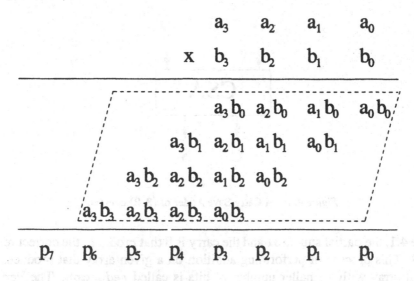

Figure 4.1. 4-bit by 4-bit Multiplication Matrix.

multiplication is concerned with not just two operands, but many of them it is imperative to organize the hardware to mitigate the carry path or chain. Therefore, many implementations consider adders according to two principles:

- Carry-Save Addition (CSA) - idea of utilizing addition without carries connected in series but just to count.

- Carry-Propagate Addition (CPA) - idea of utilizing addition with the carries connected in series to produce a result in either conventional or redundant notation.

Each adder is the same as the full adder discussed in Chapter 3, however, the view in which each connection is made from adder to adder is where the main difference lies. Because each adder is really trying to compute both carry and save information, sometimes VLSI designers refer to it as a carry-save adder or CSA. As mentioned previously, because each adder attempts to count the number of inputs that are 1, it is sometimes also called a counter. A (c, d) is an adder where c refers to the column height and d is the number of bits to display at its output. For example, a $(3, 2)$ counter counts the 3 inputs all with the same weight and displays two outputs. A $(3, 2)$ counter is shown in Figure 4.2.

Therefore, an n-bit CSA can take three n-bit operands and generate an n-bit partial sum and n-bit carry. Large operand sizes would require more CSAs to produce a result. However, a CPA would be required to produce the correct result. For example, in Table 4.1 an example is shown that adds together $A + B + D + E$ with the values $10 + 6 + 11 + 12$. The implementation utilizing the carry-save concept for this example is shown in Figure 4.4. As seen in

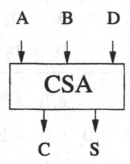

Figure 4.2. A Carry-Save Adder or $(3, 2)$ counter.

Table 4.1, the partial sum is 31 and the carry is 8 that produces the correct result of 39. This process of performing addition on a given array that produces an output array with a smaller number of bits is called *reduction*. The Verilog code for this implementation is shown in Figure 4.3.

Row	Column Value Radix 2					Column Value Radix 10
A		1	0	1	0	10
B		0	1	1	0	6
D		1	0	1	1	11
+ E		1	1	0	0	12
S	1	1	1	1	1	31
C	0	1	0	0	0	8

Table 4.1. Carry-Save Concept Example.

The CSA utilizes many topologies of adder so that the carry-out from one adder is **not** connected to the carry-in of the next adder. Eventually, a CPA could be utilized to form the true result. This organization of utilizing m-word by n-bit multi-operand adders together to add m-operands or words each of which is n-bits long is called a multi-operand adder (MOA). A m-word by n-bit multi-operand adder can be implemented using $(m - 2)$ n-bit CSA's and 1 CPA. Unfortunately, because the number of bits added together increases the result, the partial sum and carry must grow as well. Therefore, the result will contain $n + \lceil log_2(m) \rceil$ bits. In our example above, this means a $4 + log_2(4) = 6$-bit result is produced.

Higher order counters can be created by putting together various sized counters. A higher order counter (p, q) takes p input bits and produces q output bits.

```
module moa4x4 (S, C, A, B, C, D, Cin);

    input [3:0]  A, B, D, E;
    input  Cin;
    output [4:0] S, C;

    fa csa1 (c_0_0, s_0_0, A[0], B[0], D[0]);
    fa csa2 (c_0_1, s_0_1, A[1], B[1], D[1]);
    fa csa3 (c_0_2, s_0_2, A[2], B[2], D[2]);
    fa csa4 (S[4], s_0_3, A[3], B[3], D[3]);

    fa csa5 (C[0], S[0], E[0], s_0_0, Cin);
    fa csa6 (C[1], S[1], E[0], s_0_1, c_0_0);
    fa csa7 (C[2], S[2], E[0], s_0_2, c_0_1);
    fa csa8 (C[3], S[3], E[0], s_0_3, c_0,2);

endmodule // moa4x4
```

Figure 4.3. 4-operand 4-bit Multi-Operand Adder Verilog Code.

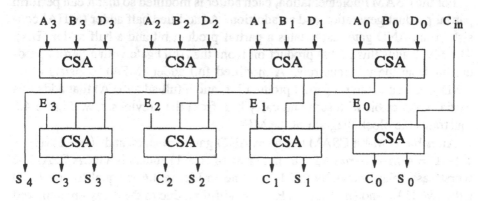

Figure 4.4. A 4 operand 4-bit Multi-Operand Adder.

Since q bits can represent a number between 0 and $2^q - 1$, the following restriction is required $p \leq 2^q - 1$. In general a $(2^q - 1, q)$ higher order counter requires $(2^q - 1 - q)$ $(3, 2)$ counters. The increase in complexity that occurs in higher-order counters and multi-operand adders can make an implementation complex as seen by the Verilog code in Figure 4.3. Consequently, some designers use programs that generate RTL code automatically. Another useful technique is to utilize careful naming methodologies for each temporary variable and declaration. For example, in Figure 4.3, the temporary variables utilize s_0_2 to represent the sum from the first carry-save adder in the second column.

4.3 Carry-Save Array Multipliers (CSAM)

The simplest of all multipliers is the carry-save array multiplier (CSAM). The basic idea behind the CSAM is that it is basically doing paper and pencil style multiplication. In other words, each partial product is being added. A 4-bit by 4-bit unsigned CSAM is shown in Figure 4.6. Each column of the multiplication matrix corresponds to a diagonal in the CSAM. The reason CSAM's are usually done in a square is because it allows metal tracks or interconnect to have less congestion. This has a tendency to have less capacitance as well as making it easier for engineers to organize the design.

The CSAM performs PP generation utilizing AND gates and uses an array of CSA's to perform reduction. The AND gates form the partial-products and the CSA's sum these partial products together or reduce them. Since most of the reduction computes the lower half of the product, the final CPA only needs to add the upper half of the product. Array multipliers are typically easy to build both using Verilog code as well in custom layout, therefore, there are many implementations that employ both.

For the CSAM implementation, each adder is modified so that it can perform partial product generation and an addition. A modified half adder (MHA) consists of an AND gate that creates a partial product bit and a half adder (HA). The MHA adds this partial product bit from the AND gate with a partial product bit from the previous row. A modified full adder (MFA) consists of an AND gate that creates a partial product bit, and a full adder (FA) that adds this partial product bit with sum and carry bits from the previous row. Figure 4.5 illustrates the block diagram of the MFA.

An n-bit by m-bit CSAM has $n \cdot m$ AND gates, m HAs, and $((n-1) \cdot (m-1)) - 1 = n \cdot m - n - m$ FAs. The final row of $(n-1)$ adders is a RCA CPA. The worst case delay shown by the dotted line in Figure 4.6 is equal to one AND gate, two HAs, and $(m+n-4)$ FAs. In addition, due to the delay encountered by each adder in the array, the worst-case delay can sometimes occur down the a_n column instead of across the diagonal. To decrease the worst case delay, the $(n-1)$-bit RCA on the bottom of the array can be replaced by a faster adder, but this increases the gate count and reduces the regularity of the design. Array multipliers typically have a complexity of $O(n^2)$ for area and $O(n)$ for delay.

The Verilog code for a 4-bit by 4-bit CSAM is shown in Figure 4.8. The hierarchy for the partial product generation is performed by the *PP* module which is shown in Figure 4.7. The MHA and MFA modules are not utilized to illustrate the hardware inside the array multiplier, however, using this nomenclature would establish a better coding structure.

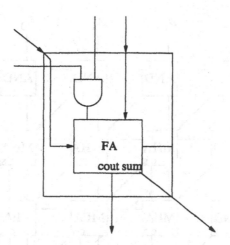

Figure 4.5. A Modified Full Adder (MFA).

4.4 Tree Multipliers

To reduce the delay of array multipliers, tree multipliers, which have $O(\log(n))$ delay, are often employed. Tree multipliers use the idea of reduction to reduce the partial products down until they are reduced enough for use with a high-speed CPA. In other words, as opposed to array multipliers, tree multipliers vary in the way that each CSA performs partial product reduction. The main objective is to reduce the partial products utilizing the carry-save concept. Each partial product is reorganized so that it can get achieve an efficient reduction array. This is possible for multiplication because each partial product in the multiplication matrix is commutative and associative with respect to addition. That is, if there are four partial products in a particular column, M, N, O, and P, it does not matter what order the hardware adds the partial products in (.e.g $M + N + O + P = P + O + N + M$).

4.4.1 Wallace Tree Multipliers

Wallace multipliers grew from an experiment into how to organize tree multipliers with the correct amount of reduction[Wal64]. Wallace multipliers group rows into sets of three. Within each three row set, FAs reduce columns with three bits and HAs reduce columns with two bits. When used in multiplier trees, full adders and half adders are often referred to as $(3, 2)$ and $(2, 2)$ counters, respectively. Rows that are not part of a three row set are transferred to the next reduction stage for subsequent reduction. The height of the matrix in the jth reduction stage is where w_j is defined by the following

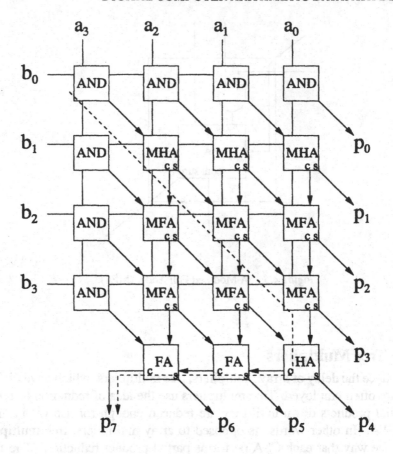

Figure 4.6. 4-bit by 4-bit Unsigned CSAM.

recursive equations [BSS01]:

$$w_0 = n$$
$$w_{j+1} = 2 \cdot \lfloor \frac{w_j}{3} \rfloor + (w_j \bmod 3)$$

To use these equations, the user first picks the bit size that they would like to design. Then, utilizing the equations above, the intermediate matrix heights are determined. For example, using Wallace's equation above, a 4-bit by 4 Wallace tree multiplier has intermediate heights of 3, and 2. Each intermediate height is the result of one level of carry-save addition (i.e. $4 \rightarrow 3 \rightarrow 2$).

The best way to visualize Wallace trees is to draw a dot diagram of the multiplication matrix. In order to draw the dot diagram, reorganize the multiplication matrix so that bits that have no space above a particular partial product column. In other words, the multiplication matrix goes from a parallelogram into a inverted triangle. For example, in Figure 4.9, a 4-bit by 4-bit dot diagram

```
module PP (P3, P2, P1, P0, X, Y);

    input [3:0]  Y;
    input [3:0]  X;

    output [3:0] P3, P2, P1, P0;

    // Partial Product Generation
    and pp1(P0[3], X[3], Y[0]);
    and pp2(P0[2], X[2], Y[0]);
    and pp3(P0[1], X[1], Y[0]);
    and pp4(P0[0], X[0], Y[0]);
    and pp5(P1[3], X[3], Y[1]);
    and pp6(P1[2], X[2], Y[1]);
    and pp7(P1[1], X[1], Y[1]);
    and pp8(P1[0], X[0], Y[1]);
    and pp9(P2[3], X[3], Y[2]);
    and pp10(P2[2], X[2], Y[2]);
    and pp11(P2[1], X[1], Y[2]);
    and pp12(P2[0], X[0], Y[2]);
    and pp13(P2[3], X[3], Y[3]);
    and pp14(P3[2], X[2], Y[3]);
    and pp15(P3[1], X[1], Y[3]);
    and pp16(P3[0], X[0], Y[3]);

endmodule // PP
```

Figure 4.7. 4-bit by 4-bit Partial Product Generation Verilog Code.

is shown. Utilizing the properties that addition is commutative and associative, the columns towards the left hand side of the multiplication matrix are reorganized upwards as shown in Figure 4.10 by the arrows.

Wallace's reduction scheme begins by grouping the rows into sets of threes. A useful technique is to draw horizontal lines in groups of threes so that its easy to visualize the reduction. In Figure 4.11 a 4-bit by 4-bit Wallace tree multiplication is shown. This multiplier takes 2 reduction stages with matrix heights of 3 and 2. The outputs of each $(3, 2)$ counters are represented as two dots connected by a diagonal line. On the other hand, the outputs of each $(2, 2)$ counter are represented as two dots connected by a crossed diagonal line. An oval is utilized to show the transition from reduction stages. In summary,

```
module array4 (Z, X, Y);

    input [3:0]  X, Y;

    output [7:0] Z;

    // Partial Product Generation
    PP pp1 (P3, P2, P1, P0, X, Y);

    // Partial Product Reduction
    ha  HA1 (carry1[2],sum1[2],P1[2],P0[3]);
    ha  HA2 (carry1[1],sum1[1],P1[1],P0[2]);
    ha  HA3 (carry1[0],sum1[0],P1[0],P0[1]);
    fa  FA1 (carry2[2],sum2[2],P2[2],P1[3],carry1[2]);
    fa  FA2 (carry2[1],sum2[1],P2[1],sum1[2],carry1[1]);
    fa  FA3 (carry2[0],sum2[0],P2[0],sum1[1],carry1[0]);
    fa  FA4 (carry3[2],sum3[2],P3[2],P2[3],carry2[2]);
    fa  FA5 (carry3[1],sum3[1],P3[1],sum2[2],carry2[1]);
    fa  FA6 (carry3[0],sum3[0],P3[0],sum2[1],carry2[0]);

    // Generate lower product bits YBITS
    buf b1(Z[0], P0[0]);
    buf b2(Z[1], sum1[0]);
    buf b3(Z[2] = sum2[0]);
    buf b4(Z[3] = sum3[0]);

    // Final Carry Propagate Addition (CPA)
    ha CPA1 (carry4[0],Z[4],carry3[0],sum3[1]);
    fa CPA2 (carry4[1],Z[5],carry3[1],carry4[0],sum3[2]);
    fa CPA3 (Z[7],Z[6],carry3[2],carry4[1],P3[3]);

endmodule // array4
```

Figure 4.8.   4-bit by 4-bit Unsigned CSAM Verilog Code.

- Dots represent partial product bits

- A uncrossed diagonal line represents the outputs of a FA

- A crossed diagonal line represents the outputs of a HA

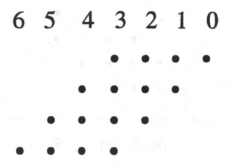

Figure 4.9. Original 4-bit by 4-bit Multiplication Matrix.

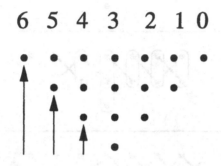

Figure 4.10. Reorganized 4-bit by 4-bit Multiplication Matrix.

This multiplier requires 16 AND gates, 6 HAs, 4 FAs and a 5-bit carry propagate adder. The total delay for the generation of the final product is the sum of one AND gate delay, one $(3, 2)$ counter delay for each of the two reduction stages, and the delay through the final carry-propagate adder. The Verilog code for the 4-bit by 4-bit Wallace tree multiplier is shown in Figure 4.12. The Verilog code utilizes an 8-bit RCA for simplicity although it could be implemented as a shorter size. In addition, as explained previously, the art of reduction can be very difficult to implement. Therefore, a good naming methodology is utilized, NX_Y_Z, where X is the reduction stage, Y is the row number for the result, and Z is the column number. For this reason, many tree multipliers are not often implemented in custom layout.

4.4.2 Dadda Tree Multipliers

Dadda multipliers are another form of Wallace multiplier, however, Dadda proposed a sequence of matrix heights that are predetermined to give the minimum number of reduction stages [Dad65]. The reduction process for a Dadda multiplier is formulated using the following recursive algorithm [BSS01], [Swa80].

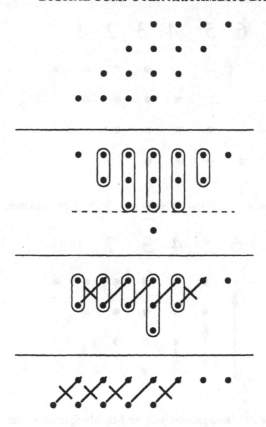

Figure 4.11.   4-bit by 4-bit Unsigned Wallace Tree Dot Diagram.

1 Let $d_1 = 2$ and $d_{j+1} = \lfloor 1.5 \cdot d_j \rfloor$, where d_j is the matrix height for the jth stage from the end. Then, proceed to find the smallest j such that at least one column of the original partial product matrix has more than d_j bits.

2 In the jth stage from the end, employ $(3, 2)$ and $(2, 2)$ counters to obtain a reduced matrix with no more than d_j bits in any column.

3 Let $j = j - 1$ and repeat step 2 until a matrix with only two rows is generated.

This method of reduction, because it attempts to compress each column, is called a column compression technique. Another advantage to utilizing Dadda multipliers is that it utilizes the minimum number of $(3, 2)$ counters [HW70]. Therefore, the number of intermediate stages is set in terms of a lower bound as:

$$2 \rightarrow 3 \rightarrow 4 \rightarrow 6 \rightarrow 9 \rightarrow \ldots$$

```
module wallace4 (Z, A, B);

    input [3:0] B;
    input [3:0] A;

    output [7:0] Z;

    // Partial Product Generation
    PP pp1 ({NO_3_6, NO_3_5, NO_3_4, NO_3_3},
    {NO_2_5, NO_2_4, NO_2_3, NO_2_2),
    {NO_1_4, NO_1_3, NO_1,2, NO_1_1},
    {NO_0_3, NO_0_2, NO_0_1, NO_0_0}, A, B);

    // Partial Product Reduction
    ha HA1(N2_1_2, N2_0_1, NO_0_1, NO_1_1);
    fa FA1(N2_1_3, N2_0_2, NO_0_2, NO_1_2, NO_2_2);
    fa FA2(N2_1_4, N2_0_3, NO_0_3, NO_1_3, NO_2_3);
    fa FA3(N2_1_5, N2_0_4, NO_1_4, NO_2_4, NO_3_4);
    ha HA2(N2_1_6, N2_0_5, NO_2_5, NO_3_5);

    ha HA3(N3_1_3, N3_0_2, N2_0_2, N2_1_2);
    fa FA4(N3_1_4, N3_0_3, N2_0_3, N2_1_3, NO_3_3);
    ha HA4(N3_1_5, N3_0_4, N2_0_4, N2_1_4);
    ha HA5(N3_1_6, N3_0_5, N2_0_5, N2_1_5);
    ha HA6(N3_1_7, N3_0_6, NO_3_6, N2_1_6);

    // Final CPA
    rca8 cpa1(carry, Cout,
            {N3_1_7, N3_0_6, N3_0_5, N3_0_4,
                N3_0_3, N3_0_2, N2_0_1, NO_0_0},
            {1'b0, N3_1_6, N3_1_5, N3_1_4,
                N3_1_3, 1'b0, 1'b0, 1'b0}, 1'b0);

endmodule // wallace4
```

Figure 4.12.   4-bit by 4-bit Unsigned Wallace Verilog Code.

In order to visualize the reductions, it is useful, as before, to draw a dotted line between each defined intermediate stage. Anything that falls below this line for a given intermediate stage, must be reduced to the given intermediate stage size or less.

The dot diagram for a 4 by 4 Dadda multiplier is shown in Figure 4.13. The Dadda tree multiplier uses 16 AND gates, 3 HAs, 3 FAs, and a 6-bit carry-propagate adder. The total delay for the generation of the final product is the sum of the one AND gate delay, one $(3, 2)$ counter delay for each of the two reduction stages, and the delay through the final 6-bit carry-propagate adder. The Verilog code for the 4-bit by 4-bit Wallace tree multiplier is shown in Figure 4.14. Again, the Verilog code utilizes an 8-bit RCA for simplicity.

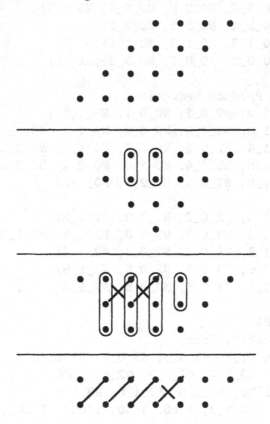

Figure 4.13.   4-bit by 4-bit Unsigned Dadda Tree Dot Diagram.

4.4.3   Reduced Area (RA) Multipliers

A recent improvement on the Wallace and Dadda reduction techniques is illustrated in a technique called Reduced Area (RA) multipliers [BSS95]. The reduction scheme of RA multipliers differ from Wallace and Dadda's methods in that the maximum number of $(3, 2)$ counters are utilized as early as pos-

```
module dadda4 (Z, A, B);

    input [3:0] B;
    input [3:0] A;

    output [7:0] Z;

    // Partial Product Generation
    PP pp1 ({NO_3_6, NO_3_5, NO_3_4, NO_3_3},
    {NO_2_5, NO_2_4, NO_2_3, NO_2_2),
    {NO_1_4, NO_1_3, NO_1,2, NO_1_1},
    {NO_0_3, NO_0_2, NO_0_1, NO_0_0}, A, B);

    // Partial Product Reduction
    ha HA1(N2_1_4, N2_0_3, NO_0_3, NO_1_3);
    ha HA2(N2_1_5, N2_0_4  NO_1_4, NO_2_4);

    ha HA3(N3_1_3, N3_0_2, NO_0_2, NO_1_2);
    fa FA1(N3_1_4, N3_0_3, NO_2_3, NO_3_3, N2_0_3);
    fa FA2(N3_1_5, N3_0_4, NO_3_4, N2_0_4, N2_1_4);
    fa FA3(N3_1_6, N3_0_5, NO_2_5, NO_3_5, N2_1_5);

    // Final CPA
    rca7 cpa1(carry, Cout,
     {NO_3_6, N3_0_5, N3_0_4, N3_0_3,
               N3_0_2, NO_0_1, NO_0_0},
     {N3_1_6, N3_1_5, N3_1_4, N3_1_3,
               NO_2_2, NO_1,1, 1'b0}, 1'b0);

endmodule // dadda4
```

Figure 4.14.   4-bit by 4-bit Unsigned Dadda Tree Verilog Code.

sible, and $(2, 2)$ counters are carefully placed to reduce the word size of the carry-propagate adder. The basic idea is to utilize a *greedier* form of Dadda's reduction scheme. In addition, since RA multipliers have fewer total dots in the reduction, they are well suited for pipelined multipliers. This is because employing $(3, 2)$ multipliers earlier than Dadda multipliers minimizes passing data between successive stages in the reduction.

The RA multiplier performs the reduction as follows [BSS95]

1 For each stage, the number of FAs used in column i is #FAs = $\lfloor b_i/3 \rfloor$, where b_i is the number of bits in column i.

2 HAs are only used

  (a) when required to reduce the number of bits in a column to the number of bits specified by the Dadda sequence.

  (b) to reduce the rightmost column containing exactly two bits.

Figure 4.15 shows the dot diagram for an 4-bit by 4-bit RA tree multiplier. This multiplier requires 16 AND gates, 3 HAs, 5 FAs, and an 4-bit carry-propagate adder. The total delay for the generation of the final product is the sum of the one AND gate delay, one $(3, 2)$ counter delay for each of the two reduction stages, and the delay through the final 4-bit carry-propagate adder. The Verilog code for the 4-bit by 4-bit Reduced Area tree multiplier is shown in Figure 4.16. Again, the Verilog code utilizes an 8-bit RCA for simplicity although it could be implemented more efficiently.

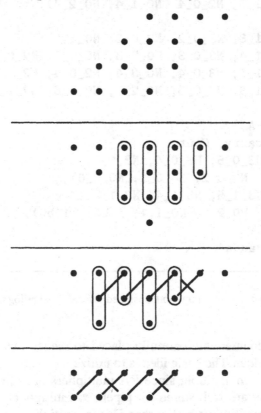

Figure 4.15.   4-bit by 4-bit Unsigned Reduced Area (RA) Diagram.

```
module ra4 (Z, A, B);

    input [3:0] B;
    input [3:0] A;

    output [7:0] Z;

    // Partial Product Generation
    PP pp1 ({NO_3_6, NO_3_5, NO_3_4, NO_3_3},
    {NO_2_5, NO_2_4, NO_2_3, NO_2_2),
    {NO_1_4, NO_1_3, NO_1,2, NO_1_1},
    {NO_0_3, NO_0_2, NO_0_1, NO_0_0}, A, B);

    // Partial Product Reduction
    ha HA1(N2_1_2, N2_0_1, NO_0_1, NO_1_1);
    fa FA1(N2_1_3, N2_0_2, NO_0_2, NO_1_2, NO_2_2);
    fa FA2(N2_1_4, N2_0_3, NO_0_3, NO_1_3, NO_2_3);
    fa FA3(N2_1_5, N2_0_4, NO_1_4, NO_2_4, NO_3_4);

    ha HA2(N3_1_3, N3_0_2, N2_0_2, N2_1_2);
    fa FA4(N3_1_4, N3_0_3, NO_3_3, N2_0_3, N2_1_3);
    ha HA3(N3_1_5, N3_0_4, N2_0_4, N2_1_4);
    fa FA5(N3_1_6, N3_0_5, NO_2_5, NO_3_5, N2_1_5);

    // Final Carry Propagate Adder
    rca7 cpa1(Z, Cout,
     {NO_3_6, N3_0_5, N3_0_4, N3_0_3,
                N3_0_2, N2_0_1, NO_0_0},
     {N3_1_6, N3_1_5, N3_1_4, N3_1_3,
                1'b0, 1'b0, 1'b0}, 1'b0);

endmodule // ra4
```

Figure 4.16. 4-bit by 4-bit Unsigned Reduced Area (RA) Verilog Code.

4.5 Truncated Multiplication

High-speed parallel multipliers are the fundamental building blocks in digital signal processing systems [MT90]. In many cases, parallel multipliers contribute significantly to the overall power dissipation of these systems [Par01].

Consequently, reducing the power dissipation of parallel multipliers is important in the design of digital signal processing systems.

In many computer systems, the $n + m$-bit products produced by the parallel multipliers are rounded to r bits to avoid growth in word size. As presented in [Lim92], truncated multiplication provides an efficient method for reducing the hardware requirements of rounded parallel multipliers. With truncated multiplication, only the $r + k$ most significant columns of the multiplication matrix are used to compute the product. The error produced by omitting the $m + n - r - k$ least significant columns and rounding the final result to r bits is estimated, and this estimate is added along with the $r + k$ most significant columns to produce the rounded product. Although this leads to additional error in the rounded product, various techniques have been developed to help limit this error [KS98], [SS93], [SD03]. This is illustrated in Figure 4.17 where a 4 by 4 truncated multiplication matrix is shown producing a 4 bit final product. The final r bits are output based on adding extra k columns and a compensation method by using a constant, variably adjusting the final result with extra bits from the partial product matrix, or combinations of both these methods.

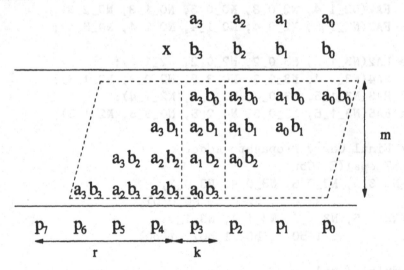

Figure 4.17. 4-bit by 4-bit Truncated Multiplication Matrix.

One method to compensate for truncation are Constant Correction Truncated (CCT) Multipliers [SS93]. In this method, a constant is added to columns $n + m - r - 1$ to $n + m - r - k$ of the multiplication matrix. The constant helps compensate for the error introduced by omitting the $n + m - r - k$ least significant columns (called reduction error), and the error due to rounding the product to r bits (called rounding error). The expected value of the sum of these error E_{total} is computed by assuming that each bit in A, B and P has an equal probability of being one or zero. Consequently, the expected value of the total error is the sum of expected reduction error and the expected rounding error as

$$E_{total} = E_{reduction} + E_{rounding}$$

$$E_{total} = \frac{1}{4}\sum_{q=0}^{S-1}(q+1)\cdot 2^{-m-n+q} + \frac{1}{2}\cdot\sum_{z=S-k}^{S-1}2^{-m-n+z}$$

where $S = m + n - r$. The constant C_{total} is obtained by rounding $-E_{total}$ to $r + k$ fractional bits, such that

$$C_{total} = -\frac{round(2^{r+k}E_{total})}{2^{r+k}}$$

where $round(x)$ indicates that x is rounded to the nearest integer.

To compute the maximum absolute error, it has been shown that the maximum absolute error occurs either when all of the partial product bits in columns 0 to $n + m - r - k - 1$ and all the product bits in columns $n + m - r - k$ to $n + m - r - k$ are ones or when they are all zeroes [SS93]. If they are all ones or all zeros, the maximum absolute error is just the constant C_{total}. Therefore, the maximum absolute error is

$$E_{max} = max(C_{total}, \sum_{q=0}^{-S-k-1}(q+1)\cdot 2^{-m-n+q} + 2^{-r}\cdot(1-2^k))$$

Although the value of k can be chosen to limit the maximum absolute error to a specific precision, this equation assumes the maximum absolute error is limited to one unit in the last place (i.e., 2^{-r}).

Figure 4.18 shows the block diagram of an $n = 4$ by $m = 4$ carry-save array CCT multiplier with $r = 4$ and $k = 2$. The rounding correction constant for the CCT array multiplier is $C_{round} = 0.0283203125$. A specialized half adder (SHA) is employed within Figure 4.18 to enable the correction constant to be added into the partial product matrix. A SHA is equivalent to a MFA that has an input set to one. The reduced half adder (RHA) and reduced full adder (RFA) are similar to an adder in that it produces only a carry without a sum since the sum is not needed. The Verilog implementation of the carry-save CCT multiplier is shown in Figure 4.19. Instead of calling a module for the partial product generation, each partial product generation logic is shown so that it is easy to see which partial products are not needed.

Another method to compensate for the truncation is using the Variable Correction Truncated (VCT) Multiplier [KS98]. Figure 4.20 shows the block diagram of an 4 by 4 carry-save array multiplier that uses the VCT Multiplication method with $r = 4$ and $k = 1$. With this type of multiplier, the values of

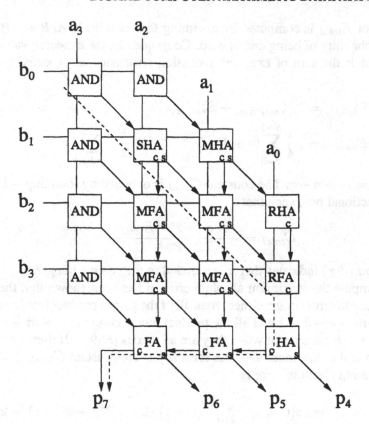

Figure 4.18. Block diagram of Carry-Save Array CCT multiplier with $n = m = r = 4$, and $k = 2$.

the partial product bits in column $m + n - r - k - 1$ are used to estimate the error due to leaving off the $m + n - r - k$ least significant columns. This is accomplished by adding the partial products bits in column $m + n - r - k - 1$ to column $m + n - r - k$. To compensate for the rounding error, a constant is added to columns $m + n - r - 2$ to $m + n - r - k$ of the multiplication matrix. The value for this constant is

$$C_{total} = 2^{-S-1}(1 - 2^{-k+1}) \tag{4.2}$$

which corresponds to the expected value of the rounding error truncated to $r + k$ bits. For the implementation in Figure 4.20, the constant is $C_{total} = 2^{4+4-4} \cdot (1 - 1) = 0.0$, therefore, there is no need to add a SHA in Figure 4.20. The Verilog implementation of the carry-save VCT multiplier is shown in Figure 4.21.

When truncation occurs, the diagonals that produce the $t = m + n - r - k$ least significant product bits are eliminated. To compensate for this, the

```
// Correction constant value: 0.0283203125 (000010)
module array4c (Z, X, Y);

    input [3:0]  Y;
    input [3:0]  X;

    output [3:0] Z;

    // Partial Product Generation
    and pp1(P0[3], X[3], Y[0]);
    and pp2(P0[2], X[2], Y[0]);
    and pp3(sum1[3], X[3], Y[1]);
    and pp4(P1[2], X[2], Y[1]);
    and pp5(P1[1], X[1], Y[1]);
    and pp6(sum2[3], X[3], Y[2]);
    and pp7(P2[2], X[2], Y[2]);
    and pp8(P2[1], X[1], Y[2]);
    and pp9(P2[0], X[0], Y[2]);
    and pp10(sum3[3], X[3], Y[3]);
    and pp11(P3[2], X[2], Y[3]);
    and pp12(P3[1], X[1], Y[3]);
    and pp13(P3[0], X[0], Y[3]);

    // Partial Product Reduction
    specialized_half_adder SHA1(carry1[2],sum1[2],P1[2],
                                P0[3]);
    ha  HA1(carry1[1],sum1[1],P1[1],P0[2]);
    fa  FA1(carry2[2],sum2[2],P2[2],sum1[3],carry1[2]);
    fa  FA2(carry2[1],sum2[1],P2[1],sum1[2],carry1[1]);
    assign carry2[0] = P2[0] & sum1[1];
    fa  FA3(carry3[2],sum3[2],P3[2],sum2[3],carry2[2]);
    fa FA4(carry3[1],sum3[1],P3[1],sum2[2],carry2[1]);
    reduced_full_adder  RFA1(carry3[0],P3[0],sum2[1],
                             carry2[0]);

    // Final Carry Propagate Addition
    ha CPA1(carry4[0],Z[0],carry3[0],sum3[1]);
    fa CPA2(carry4[1],Z[1],carry3[1],carry4[0],sum3[2]);
    fa CPA3(Z[3],Z[2],carry3[2],carry4[1],sum3[3]);

endmodule // array4c
```

Figure 4.19. Carry-Save Array CCT multiplier with $n = m = r = 4$, and $k = 2$ Verilog Code.

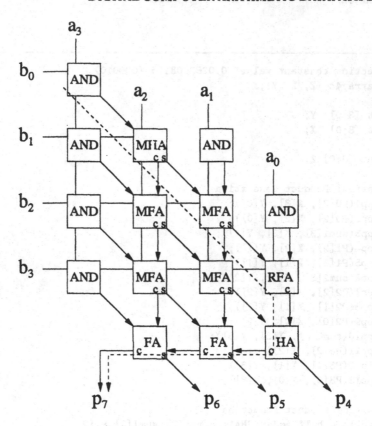

Figure 4.20. Block Diagram of carry-save array VCT multiplier with $n = m = r = 4$, and $k = 1$.

AND gates that generate the partial products for column $t - 1$ are used as inputs to the modified adders in column t. As explained previously, since the k remaining modified full adders on the right-hand-side of the array do not need to produce product bits, they are replaced by modified reduced half and full adders (RFAs), which produce a carry, but do not produce.

Another method for truncation, called a Hybrid Correction Truncated (HCT) Multiplier [SD03], uses both constant and variable correction techniques to reduce the overall error. In order to implement a HCT multiplier, a new parameter is introduced, p, that represents the percentage of variable correction to use for the correction. This percentage is utilized to chose the number of partial products from column $m + n - r - k - 1$ to be used to add into column $m + n - r - k$. The calculation of the number of variable correction bits is the following utilizing the number of bits used in the variable correction method, $N_{variable}$

$$N_{variable_{hybrid}} = floor(N_{variable} \times p) \qquad (4.3)$$

```
// Correction constant value: 0.0
module array4v (Z, X, Y);

    input [3:0]  Y;
    input [3:0]  X;

    output [3:0] Z;

    // Partial Product Generation
    and pp1(P0[3], X[3], Y[0]);
    and pp2(sum1[3], X[3], Y[1]);
    and pp3(P1[2], X[2], Y[1]);
    and pp4(carry1[1], X[1], Y[1]);
    and pp5(sum2[3], X[3], Y[2]);
    and pp6(P2[2], X[2], Y[2]);
    and pp7(P2[1], X[1], Y[2]);
    and pp8(carry2[0], X[0], Y[2]);
    and pp9(sum3[3], X[3], Y[3]);
    and pp10(P3[2], X[2], Y[3]);
    and pp11(P3[1], X[1], Y[3]);
    and pp12(P3[0], X[0], Y[3]);

    // Partial Product Reduction
    ha  HA1(carry1[2],sum1[2],P1[2],P0[3]);
    fa  FA1(carry2[2],sum2[2],P2[2],sum1[3],carry1[2]);
    fa  FA2(carry2[1],sum2[1],P2[1],sum1[2],carry1[1]);
    fa  FA3(carry3[2],sum3[2],P3[2],sum2[3],carry2[2]);
    fa  FA4(carry3[1],sum3[1],P3[1],sum2[2],carry2[1]);
    reduced_full_adder  RFA1(carry3[0],P3[0],sum2[1],
                            carry2[0]);

    // Final Carry Propagate Addition
    ha  CPA1(carry4[0],Z[0],carry3[0],sum3[1]);
    fa  CPA2(carry4[1],Z[1],carry3[1],carry4[0],sum3[2]);
    fa  CPA3(Z[3],Z[2],carry3[2],carry4[1],sum3[3]);

endmodule // array4v
```

Figure 4.21. Carry-Save Array VCT multiplier with $n = m = r = 4$, and $k = 1$ Verilog Code.

Similar to both the CCT and the VCT multipliers, a HCT multiplier uses a correction constant to compensate for the rounding error. However, since the correction constant will be based on a smaller number bits than a VCT multiplier, the correction constant is modified as follows

$$C_{VCT'} = 2^{-r-k-2} \cdot N_{variable_{hybrid}} \tag{4.4}$$

This produces a new correction constant based on the difference between the new variable correction constant and the constant correction constant:

$$C_{round} = \frac{round((C_{CCT} - C_{VCT'}) \cdot 2^{r+k})}{2^{r+k}} \tag{4.5}$$

Figure 4.22 shows an a 4 by 4 carry-save array multiplier that uses the HCT Multiplication method with $r = 4$, $k = 1$, and $p = 0.4$. The rounding correction constant for the HCT array multiplier is $C_{round} = 0.0244140625$, which is implemented in the block diagram by changing one of the MHAs in the second row to a SHA. Since the HCT multiplier is being compensated by utilizing constant correction **and** variable correction, the constant is $C_{VCT} \le C_{HCT} \le C_{CCT}$. The Verilog implementation of the carry-save HCT multiplier is shown in Figure 4.23 For all the multipliers presented here, a similar modification can be performed for tree multipliers, such as Dadda multipliers. Truncated multipliers can also be combined with non-truncated parallel multipliers [WSS01].

4.6 Two's Complement Multiplication

Most computing systems involve the use of signed and unsigned binary numbers. Therefore, multiplication requires some mechanism to compute two's complement multiplication. The most common implementation for two's complement multipliers is to use the basic mathematical equation for integer or fractional multiplication and use algebra to formalize a structure. The most popular of these implementation are called Baugh-Wooley [BW73] or Pezaris [Pez71] multipliers named after each individual who presented them. Therefore, in this section we formalize two's complement multiplication the same way.

Although previously we have introduced fractional binary numbers, the formation of the new multiplication matrix is easiest seen utilizing two's complement binary integers. Conversion to an n-bit binary fraction from a n-bit binary integer is easily accomplished by dividing by 2^{n-1}. For two's complement binary integers, a number can be represented as:

$$A = -a_{n-1} \cdot 2^{n-1} + \sum_{i=0}^{n-2} a_i \cdot 2^i$$

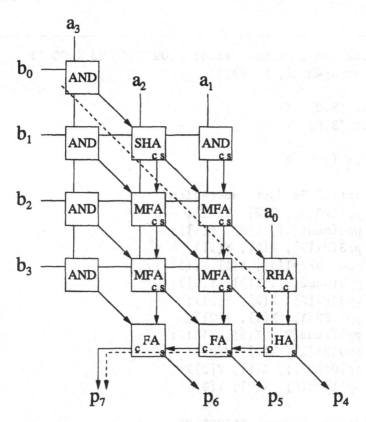

Figure 4.22. Block diagram of Carry-Save Array HCT multiplier with $n = m = r = 4$, $k = 1$, and $p = 0.4$.

Consequently, the multiplication of two n-bit two's complement binary integers A and B creates the product P with the value as follows

$$
\begin{aligned}
P &= A \cdot B \\
&= \left[-a_{n-1} \cdot 2^{n-1} + \sum_{i=0}^{n-2} a_i \cdot 2^i \right] \cdot \left[-b_{n-1} \cdot 2^{n-1} + \sum_{j=0}^{n-2} b_j \cdot 2^j \right] \\
&= a_{n-1} \cdot b_{n-1} \cdot 2^{2 \cdot n - 2} + \sum_{i=0}^{n-2} \sum_{j=0}^{n-2} a_i \cdot b_j \cdot 2^{i+j} \\
&\quad - \sum_{i=0}^{n-2} a_i \cdot b_{n-1} \cdot 2^{i+n-1} - \sum_{j=0}^{n-2} a_{n-1} \cdot b_j \cdot 2^{j+n-1}
\end{aligned}
$$

From this algebraic manipulation it is apparent that two of the distributed terms of the multiplication are negative. One of the ways to implement this

```
// Correction constant value: 0.0244140625 (00001)
module array4h (Z, X, Y);

    input [3:0]  Y;
    input [3:0]  X;

    output [3:0] Z;

    // Partial Product Generation
    and pp1(P0[3], X[3], Y[0]);
    and pp2(sum1[3], X[3], Y[1]);
    and pp3(P1[2], X[2], Y[1]);
    and pp4(carry1[1], X[1], Y[1]);
    and pp5(sum2[3], X[3], Y[2]);
    and pp6(P2[2], X[2], Y[2]);
    and pp7(P2[1], X[1], Y[2]);
    and pp8(sum3[3], X[3], Y[3]);
    and pp9(P3[2], X[2], Y[3]);
    and pp10(P3[1], X[1], Y[3]);
    and pp11(P3[0], X[0], Y[3]);

    // Partial Product Reduction
    specialized_half_adder  SHA1(carry1[2],sum1[2],P1[2],
                                 P0[3]);
    fa  FA1(carry2[2],sum2[2],P2[2],sum1[3],carry1[2]);
    fa  FA2(carry2[1],sum2[1],P2[1],sum1[2],carry1[1]);
    fa  FA3(carry3[2],sum3[2],P3[2],sum2[3],carry2[2]);
    fa  FA4(carry3[1],sum3[1],P3[1],sum2[2],carry2[1]);
    assign carry3[0] = P3[0] & sum2[1];

    // Final Carry Propagate Addition
    ha CPA1(carry4[0],Z[0],carry3[0],sum3[1]);
    fa CPA2(carry4[1],Z[1],carry3[1],carry4[0],sum3[2]);
    fa CPA3(Z[3],Z[2],carry3[2],carry4[1],sum3[3]);

endmodule // array4h
```

Figure 4.23. Carry-Save Array HCT multiplier with $n = m = r = 4$, $k = 1$, and $p = 0.4$ Verilog Code.

in digital arithmetic is to convert the negative value into a two's complement number. This is commonly done by taking the one's complement of the number and adding 1 to the unit in the least significant position (*ulp*). This is referred to as adding an *ulp* as opposed to a 1 because it guarantees the conversion regardless of the location of the radix point. Therefore, the equations above can be manipulated one more time as shown below. The constants occur because the one's complement operation for the two terms in column $2^{2 \cdot n - 2}$ (since its originally a 0 and gets complemented to a 1) and the ulp in column 2^{n-1}.

$$P = a_{n-1} \cdot b_{n-1} \cdot 2^{2 \cdot n - 2} + \sum_{i=0}^{n-2} \sum_{j=0}^{n-2} a_i \cdot b_j \cdot 2^{i+j}$$

$$+ \sum_{i=0}^{n-2} \overline{a_i \cdot b_{n-1}} \cdot 2^{i+n-1} + \sum_{j=0}^{n-2} \overline{a_{n-1} \cdot b_j} \cdot 2^{j+n-1}$$

$$+ (2^{2 \cdot n - 2} + 2^{-2 \cdot n - 2}) + (2^{n-1} + 2^{n-1})$$

$$P = a_{n-1} \cdot b_{n-1} \cdot 2^{2 \cdot n - 2} + \sum_{i=0}^{n-2} \sum_{j=0}^{n-2} a_i \cdot b_j \cdot 2^{i+j}$$

$$+ \sum_{i=0}^{n-2} \overline{a_i \cdot b_{n-1}} \cdot 2^{i+n-1} + \sum_{j=0}^{n-2} \overline{a_{n-1} \cdot b_j} \cdot 2^{j+n-1}$$

$$+ 2^{2 \cdot n - 1} + 2^n$$

From the equations listed above, a designer can easily implement the design utilized previous implementations such as the carry-save array multiplier. matrix, except $2 \cdot n - 2$ partial products are inverted and ones are added in columns n and $2 \cdot n - 1$. Figure 4.24 shows a 4-bit by 4-bit two's complement carry-save array multiplier. The $2 \cdot n - 2 = 6$ partial products are inverted by changing $2 \cdot n - 2 = 6$ AND gates to NAND gates. Negating MFAs (NMFAs) are similar to the MFA, except that the AND gate is replaced by a NAND gate. A specialized half adder (SHA) adds the one in column n with sum and carry vectors from the previous row as in the truncated multiplier. It implements the equations

$$s_i = a_i \oplus b_i \oplus 1 = \overline{a_i \oplus b_i}$$
$$c_{i+1} = (a_i \cdot b_i) + (a_i + b_i) \cdot 1 = a_i + b_i$$

The inverter that complements the $2 \cdot n - 1$ column adds the 1 in column $2 \cdot n - 1$ since column $2 \cdot n - 1$ only requires a half adder and since one of the input operands is a 1 (i.e. $sum = a \oplus 1 = \overline{a}$). The Verilog code for the 4-bit by 4-bit two's complement carry-save multiplier is shown in Figure 4.26. Since the partial products require the use of 6 NAND gates, Figure 4.25 is shown to illustrate the modified partial product generation. A similar modification can be performed for tree multipliers such as Dadda multipliers.

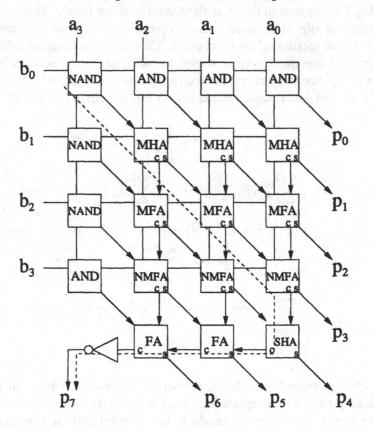

Figure 4.24.   4-bit by 4-bit Two's Complement CSAM.

4.7   Signed-Digit Numbers

Another popular method to handle negative numbers is the use of the Signed-Digit (SD) number system [Avi61]. Signed-Digit (SD) number systems allow both positive and negative digits. The range of digits is

$$a_i \in \{\overline{\alpha}, \overline{\alpha} + 1, \dots \overline{1}, 0, 1, \dots, \alpha - 1, \alpha\}$$

where $\overline{x} = -x$ and $\lceil \frac{r-1}{2} \rceil \leq \alpha \leq r - 1$. Since a SD number system can represent more than one number, it is typically called a redundant number system. For example, the value 5 in radix 10 can be represented in radix 2 or binary as 0101 or 011$\overline{1}$. Utilizing redundancy within an arithmetic datapath can have certain advantages since decoding values can be simplified which will be ex-

```
module PPtc (P3, P2, P1, P0, X, Y);

   input [3:0]  Y;
   input [3:0]  X;

   output [3:0] P3, P2, P1, P0;

   // Partial Product Generation
   nand pp1(P0[3], X[3], Y[0]);
   and pp2(P0[2], X[2], Y[0]);
   and pp3(P0[1], X[1], Y[0]);
   and pp4(P0[0], X[0], Y[0]);
   nand pp5(P1[3], X[3], Y[1]);
   and pp6(P1[2], X[2], Y[1]);
   and pp7(P1[1], X[1], Y[1]);
   and pp8(P1[0], X[0], Y[1]);
   nand pp9(P2[3], X[3], Y[2]);
   and pp10(P2[2], X[2], Y[2]);
   and pp11(P2[1], X[1], Y[2]);
   and pp12(P2[0], X[0], Y[2]);
   and pp13(P2[3], X[3], Y[3]);
   nand pp14(P3[2], X[2], Y[3]);
   nand pp15(P3[1], X[1], Y[3]);
   nand pp16(P3[0], X[0], Y[3]);

endmodule // PP
```

Figure 4.25.   4-bit by 4-bit Partial Product Generation for Two's Complement Verilog Code.

plored in later designs. A useful term that measures the amount of redundancy is $\rho = \frac{a}{r-1}$ where $\rho > 1/2$.

To convert the value of a n-bit, radix-r SD integer, the following equation can be utilized:

$$A = \sum_{i=0}^{n-1} a_i \cdot r^i$$

For example, $A = 1\bar{1}0\bar{1}$ in radix 2 has the value

$$A = -1 \cdot 2^0 + 0 \cdot 2^1 + -1 \cdot 2^2 + 1 \cdot 2^3 = 3_{10}$$

```
module array4tc (Z, X, Y);

    input [3:0]  X, Y;

    output [7:0] Z;

    // Partial Product Generation
    PPtc pptc1 (P3, P2, P1, P0, X, Y);

    // Partial Product Reduction
    ha   HA1 (carry1[2],sum1[2],P1[2],P0[3]);
    ha   HA2 (carry1[1],sum1[1],P1[1],P0[2]);
    ha   HA3 (carry1[0],sum1[0],P1[0],P0[1]);
    fa   FA1 (carry2[2],sum2[2],P2[2],P1[3],carry1[2]);
    fa   FA2 (carry2[1],sum2[1],P2[1],sum1[2],carry1[1]);
    fa   FA3 (carry2[0],sum2[0],P2[0],sum1[1],carry1[0]);
    fa   FA4 (carry3[2],sum3[2],P3[2],P2[3],carry2[2]);
    fa   FA5 (carry3[1],sum3[1],P3[1],sum2[2],carry2[1]);
    fa   FA6 (carry3[0],sum3[0],P3[0],sum2[1],carry2[0]);

    // Generate lower product bits YBITS
    buf b1(Z[0], P0[0]);
    buf b2(Z[1], sum1[0]);
    buf b3(Z[2] = sum2[0]);
    buf b4(Z[3] = sum3[0]);

    // Final Carry Propagate Addition (CPA)
    ha CPA1 (carry4[0],Z[4],carry3[0],sum3[1]);
    fa CPA2 (carry4[1],Z[5],carry3[1],carry4[0],sum3[2]);
    fa CPA3 (cout,Z[6],carry3[2],carry4[1],P3[3]);
    not i1 (Z[7], cout);

endmodule // array4tc
```

Figure 4.26.   4-bit by 4-bit Signed CSAM Verilog Code.

Consequently, a carry-propagate adder is required to convert a number in SD number system to a conventional number system. Another convenient method for converting a number in SD number to a conventional number is by subtracting the negative weights from the digits with positive weights. Other methods have also been proposed for converting redundant number systems to conventional binary number systems [SP92],[YLCL92]. For example, if $A = 3\overline{2}10\overline{3}$ for radix 10, we obtain

$$
\begin{array}{ccccc}
3 & 0 & 1 & 0 & 0 \\
0 & 2 & 0 & 0 & 3 \\
\hline
2 & 8 & 0 & 9 & 7
\end{array}
$$

Table 4.2. SD Conversion to a Conventional Binary Number System.

Another potential advantage for the SD number system is that it can be utilized within adders so that addition is carry-free. Theoretically, a SD adder implementation can be independent of the length of the operands. Unfortunately, utilizing a SD number system might be costly since a larger number of bits are required to represent each bit. Moreover, the SD adder has to be designed in such a way a new carry will be not be generated, such that $\mid sum_i \mid \leq a$, imposing a larger constraint on the circuit implementation.

However, one of the biggest advantages for the SD number systems is that they can represent numbers with the smallest number of non-zero digits. This representation is sometimes known as a canonical SD (CSD) representation or minimal SD representations. For example, 00111111 is better utilized with hardware as $0100000\overline{1}$. CSD representations never have two adjacent non-zero digits thus simplifying the circuit implementation even further. Extended information regarding SD notation and its implementation can be found in [EL03], [Kor93].

As mentioned previously, tree multipliers have irregular structures. The irregularity complicates the implementation. Therefore, most implementations of tree multipliers typically resort to standard-cell design flows. Like the carry-save adder, a SD adder can generate the sum of two operands in constant time independent of the size of the operands. On the other hand, SD adder trees typically resorts in a nice regular structure which makes it a great candidate for custom layout implementations. Unfortunately, the major disadvantage of the SD adder is that its design is more complex resorting in an implementation that consumes more area.

Another structure, called a compressor, is potentially advantageous for conventional implementations. A (p, q) compressor takes p inputs and produces q outputs. In addition, it takes k carry-in bits and produces k carry-out bits. The $(4, 2)$ compressor, which is probably the most common type of compressor, takes 4 input bits and 1 carry-in bit, and produces 2 output bits and 1 carry-out bit [Wei82]. The fundamental difference between compressors and multi-operand adders is that the carry-out bit does not depend on the carry-in

bit. Although compressor and SD trees can be beneficial for the implementations listed here, they are not explored since the basic idea of multipliers is the
general interest of this text.

4.8   Booth's algorithm

Booth's algorithms can also be used to convert from binary representations
to SD representations [Boo51]. However, the representations produced are not
guaranteed to be canonical.   With the radix 2 Booth algorithm, groups of
two binary digits, b_i and b_{i-1} are used to determine the binary signed digit
d_i, according to the Table 4.3. Booth's algorithm is based on the premise of
SD notation in that fewer partial products have to be generated for groups of
consecutive zeroes and ones in the multiplier. In addition, for multiplier bits
that have consecutive zeroes there is no need to generate any partial product
which improves overall performance for timing. In other words, only a shift is
required for every 0 in the multiplier.

Booth's algorithm utilizes the advantage of SD notation by incorporating
the conversion to conventional binary notation within the structure. In other
words, numbers like $\dots 011 \dots 110 \dots$ can be changed to $\dots 100 \dots 0\bar{1}0 \dots$.
Therefore, instead of generating all m partial products, only two partial products need to be generated. To invoke this in hardware, the first partial product is
added, whereas, the second is subtracted. This is typically called *recoding* in
SD notation [Par90]. Although there are many different encodings for Booth,
the original Booth encoding is shown in Table 4.3. For this algorithm, the current bit b_i and the preceding bit b_{i-1} of the multiplier are examined to generate
the $i$the bit of the recoded multiplier. For $i = 0$, the preceding bit x_{-1} is set to
0. In summary, the recoded bit can be computed as:

$$d_i = b_{i-1} - b_i$$

b_i	b_{i-1}	d_i	Comment
0	0	0	String of zeros
0	1	1	End of a string of ones
1	0	$\bar{1}$	Start of a string of ones
1	1	0	String of ones

Table 4.3.   Radix-2 Booth Encoding.

Booth's algorithm can also be modified to handle two's complement numbers, however, special attention is required for the sign bit. Similar to other

operations, the sign bit is examined to determine if an addition or subtraction is required. However, since the sign bit only determines the sign, no shift operation is required. In this examination, we will show two different implementations of Booth for unsigned and signed arithmetic. Many implementations inadvertently assume that Booth always handles signed arithmetic, however, this is not totally true. Therefore, specialized logic must be incorporated into the implementation.

One approach for implementing Booth's algorithm is to put the implementation into a carry-save array multiplier [MK71]. This multiplier basically involves n rows where n is the size of the multiplier. Each row is capable of adding or adding and shifting the output to the proper place in the multiplication matrix so it can be reduced. The basic cell in this multiplier is the controlled add, subtract, and shift (CAS) block. The same methodology is utilized as the carry-save array multiplier where a single line without an arrow indicates the variables pass through the CAS cell into the next device. The equations to implement the CAS cell are the following:

$$s_{out} = s_{in} \oplus (a \cdot H) \oplus (c_{in} \cdot H)$$
$$c_{out} = (s_{in} \oplus D) \cdot (a + c_{in}) + (a \cdot c_{in})$$

The values of H and D come from the CTRL block which implements the recoding in Table 4.3. The two control signals basically indicate to the CAS cell whether a shift and add or subtract are performed. if $H = 0$, then a shift is only required and $s_{out} = s_{in}$. On the other hand, if $H = 1$, then a full adder is required where a is the multiplicand bit. However, for a proper use of the SD notation, a $\bar{1}$ indicates that a subtraction must occur. Therefore, if $D = 0$, a normal carry out signal is generated, whereas, if $D = 1$ a borrow needs to be generated for the proper sum to be computed.

4.8.1 Bitwise Operators

The equations above are typically simple operators to implement in Verilog as RTL elements as shown in previous chapters. Sometimes a designer may not want to spend time implementing logic gate per gate, but still wants a structural implementation. One way to accomplish this is with bitwise operators. Bitwise operators, shown in Table 4.4, operate on the bits of the operand or operands. For example, the result of $A\&B$ is the AND of each corresponding bit of A with B. It is also important to point out that these operators match with BLIF output [oCB92] produced by programs such as ESPRESSO and SIS [SSL$^+$92]. This makes the operator and the use of the $assign$ keyword simple for any logic. However, it is should be **strongly** pointed out that this may lead to a behavioral level implementation. On the other hand, for two-level logic and small multi-level logic implementations, this methodology for implementing logic in Verilog is encouraged.

Operator	Operation
~	Bitwise negation
&	Bitwise AND
|	Bitwise OR
^	Bitwise XOR
~&	Bitwise NAND
~|	Bitwise NOR

Table 4.4. Bitwise Operators.

Therefore, it is easy to implement the equations for the CTRL and CAS cells with these new operators and the *assign* keyword. The naming convention for each Verilog module is still implemented as before. However, instead of inserting instantiations for each gate, one line is required preceded with an *assign* keyword. For example, the CAS module can be implemented as follows:

```
assign W = B ^ D;
assign sout = B ^ (a & H) ^ (cin & H);
assign cout = (W & (a | cin)) | (a & cin);
buf b1(U, a);
buf b2(Q, H);
buf b3(R, D);
```

The *buf* statements are inserted to allow the single lines without an arrow to pass the variables to the next CAS cell.

```
assign H = x ^ x_1;
assign D = x & (~x_1);
```

The block diagram of the carry-save radix 2 Booth multiplier is in Figure 4.27. The first row corresponds to the most significant bit of the multiplier. The partial products generated in this row need to be shifted to the left before its added or subtracted to a multiple of the multiplicand. Therefore, one block is added to the end of the row and each subsequent row.

It is also interesting to note that the left most bit of the multiple of the multiplicand is replicated to handle the sign bit of the multiplicand. Although the implementation listed here implements Booth's algorithm, it does not take advantage of the strings of zeroes or ones. Therefore, rows can not be eliminated. However, this design does incorporate the correction step above (i.e. the subtractions in Section 4.6) into the array so that it can handle two's complement numbers. In the subsequent section, implementations for Booth will be shown that are for both unsigned and signed numbers.

The Verilog implementation is shown in Figure 4.28. Since each CAS cell can be complicated by the multiple input and output requirements, a comment is placed within the Verilog code to help the user implement the logic. This comment is basically a replica of the module for the CAS cell. Having this comment available helps the designer guide the inputs and outputs to their proper port. A mistake in ordering the logic could potentially cause the Verilog code to compile without any potential error message leading to many hours of debugging. In addition, the *assign* statement is also utilized to add a constant *GND*.

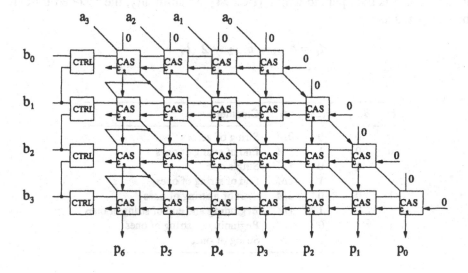

Figure 4.27. 4-bit by 4-bit Signed Booth's Carry Save Array Multiplier (Adapted from [Kor93]).

4.9   Radix-4 Modified Booth Multipliers

A disadvantage to Booth's algorithm presented in the previous section is that the algorithm can become inefficient when zeroes and ones are interspersed randomly within a given input. This can be improved by examining three bits of the multiplier at a time instead of two, [Bew94], [Mac61]. This obviously reduces the number of partial products by half and is called radix 4 *Modified Booth's Algorithm*. In this section, we examine a different implementation of Booth multipliers. Instead of utilizing a controlled adder or subtractor, a multiplexor is utilized to choose a 0 or a multiple of the multiplicand. The output of the multiplexor can then be fed into an adder tree. This is the most common type of organization utilized today for Booth multipliers because it facilitates the logic into a Booth decoder and a Booth selector.

Radix 4 Booth encoded digits have values from $d_i \in \{\bar{2}, \bar{1}, 0, 1, 2\}$. Radix 4 digits can also be obtained directly from two's complement values, according to Table 4.5. In this case, groups of three bits are examined, with one bit overlap between groups. The general idea is to reduced the number of partial products by grouping the bits of the multiplier into pairs and selecting the partial products from the set $\{0, M, 2M\}$ where M is the multiplicand. Because this implementation typically outputs 2 bits for decoding the bits, it sometimes is called a *Booth* 2 multiplier. Each partial product is shifted two bit positions with respect to the previous row. In general, there will be $\lfloor \frac{n+2}{2} \rfloor$ partial products where n is the operand length [Bew94]. In summary, the recoded bit can be computed as:

$$d_i = b_{i-1} + b_{i-2} - 2 \cdot b_i$$

b_i	b_{i-1}	b_{i-2}	d_i	Comment
0	0	0	$0M$	String of zeroes
0	0	1	$1M$	End of string of ones
0	1	0	$1M$	Single one
0	1	1	$2M$	End of string of ones
1	0	0	$\bar{2}M$	Start of a string of ones
1	0	1	$\bar{1}M$	Beginning and end of string of ones
1	1	0	$\bar{1}M$	Beginning of string of ones
1	1	1	$\bar{0}M$	String of ones

Table 4.5.   Radix-4 Booth Encoding.

One of the interesting components to Booth multipliers is that it employs sign extension to make sure that the sign is propagated appropriately. Since SD notation is utilized, the sign of the most significant bit must be sign extended to allow the proper result to occur. For Booth multipliers, because SD notation is involved, sign extension is required for both unsigned and signed logic. Figure 4.29 shows an implementation 4-bit by 4-bit unsigned radix-4 modified Booth multiplier block diagram illustrating how sign extension is utilized. The partial products are represented as dots as in tree multiplier. Each partial product, except for the bottom one, is 5 bits since numbers as large as two times the multiplicand should be handled. The bottom row is only 4 bits since the multiplier is padded with 2 zeroes to guarantee a positive result.

Each level in Figure 4.29 is given a letter (a) through (d) showing how sign extension is implemented. In (a), the original multiplication matrix is shown assuming that the partial products bits are negative causing ones to be sign extended. In (b), the ones are simplified assuming the most significant column ones are added together. This is the finalized multiplication matrix. Assuming a designer would like logic to implement the matrix assuming the multiplicand could be positive or negative, (c) is shown where $S1$ represents the sign bit for row 1. This logic allows the sign extension to occur if $S1 = 1$ for row 1 and also for $S2$. Since this implementation is for unsigned numbers, row 3 does not require sign extension. Finally, (d) shows the final simplified multiplication matrix similar to (b) for the 4-bit by 4-bit unsigned radix-4 modified Booth multiplier.

The Verilog code for the Booth 2 decoder is shown in Figure 4.30. The values $M1$ and $M2$ are the two outputs for $+M$ and $-2M$, respectively. The Booth 2 selector, shown in Figure 4.31, selects the proper shifted value of the Multiplicand. In addition, the exclusive-or gates allow a negative value of the multiplicand to be generated. The multiplexor, $mux21h$, in the selector employs a *one-hot* type of encoding. One hot encoding is typically utilized to simplify the logic. The equations for a 2-input one-hot multiplexor is as follows:

$$Z = A \cdot S1 + B \cdot S2$$

This multiplexor has two inputs A and B which are chosen based on the two select signals $S1$ and $S2$ and outputted as Z. In one-hot encoding only one of the selecting bits is on at one time. Therefore, having $S1 = S2 = 1$ is not a valid encoding. The finalized 4-bit by 4-bit unsigned radix-4 modified Booth multiplier is shown in Figure 4.32. In this code, multi-operand adders are utilized. A final carry-propagate adder is utilized to complete the product. The Verilog implementation of the one-hot multiplexor is not shown since it could easily be implemented using bitwise operators.

4.9.1 Signed Radix-4 Modified Booth Multiplication

It is easy to modify the previous multiplication matrix to handle signed multiplication. Since the most significant product is necessary to guarantee a positive result, it is not needed for signed multiplication. The new multiplication matrix is shown in Figure 4.33. There are two additional modifications that are necessary. First, when $\pm M$ is chosen from Table 4.5 (i.e. entries 1, 2, 5, or 6 from the partial product selection table), an empty bit is seen in the most significant bit of the Booth selector. In the unsigned version, a 0 is placed in this slot for the $mux1$ instantiation. However, for the signed version, sign extension must occur as shown in Figure 4.34.

Second, the most significant modification to the unsigned matrix is that the sign extension is not that straight forward. The leading ones for a particular partial product are cleared when that partial product is positive. For signed multiplication, this occurs when the multiplicand is positive and the multiplier select bits chooses a positive multiple. It also occurs when the multiplicand is negative and the multiplier select bits choose a negative multiplier. This is implemented as a simple exclusive NOR (XNOR) between the sign bit of the multiplicand and the most significant bit of the partial product selection bit (i.e. remember this is sign extended for signed arithmetic). This bit is called $E1$ in Figure 4.33. The complement $\overline{E1}$ is required to automatically choose between a signed and unsigned multiplicand. As in unsigned multiplication, each level in also shown in Figure 4.33 showing how sign extension is implemented. The reference letters (a) through (d) utilize the same description as in the unsigned multiplication with the exceptions noted above.

The Verilog code for the Booth 2 signed multiplication is shown in Figure 4.35. The decoder is not shown since it the same as shown in Figure 4.30. The xn instantiation in Figure 4.34 implements the XNOR and XOR for EX and $\overline{EX}$ where X denotes the row.

4.10   Fractional Multiplication

Many digital signal processors (DSPs) and embedded processors compute utilizing fixed-point arithmetic [EB00]. Because of this, a correct examination of the radix point is necessary when multiplication is involved. For example, Figure 4.36 shows the unsigned multiplication matrix of $X = 10.01 = 2.25$ by $Y = 1.011 = 1.375$ and $P = 011.00011 = 3.09375$. The multiplicand has two integer bits and two fractional bits, whereas, the multiplier has one integer bit and three fractional bits. Before any design is implemented, it is important to determine the precision that a computation has either through error analysis or by its range.

This becomes increasingly important when two's complement multiplication is involved. In certain circumstances, several sign bits may be produced. For example, suppose that two's complement multiplication is performed in the example above assuming that the range of X is $1.75 \geq| X |$ and the range of Y is $0.875 \geq| Y |$. Since two's complement numbers have non symmetrical ranges, this example assumes that -2.0 and -1.0 for X and Y are not possible values, respectively. This means that the range of the product is $1.53125 \geq| P |$. Since this only requires one integer bit, the other two integer bits are sign bits. If the product was being sent to another logic block the integer bit, $P[5]$ and $P[6]$ **or** $P[7]$ could be utilized in this new logic block since $P[6]$ **and** $P[7]$ are sign bits.

Therefore, when implementing logic with fractional and integer bits it is important to be aware of the range of the product. The range of the operands does not always set the range of the product. A specific algorithm, such as the Newton-Raphson iteration seen in Chapter 7, can limit the range smaller than a given precision. DSPs tend to exploit different fractional notation since many applications require formats in different precisions [FG00], [Ses98]. Fixed-point DSPs adopt the notation that arithmetic is $(S.F)$ where S represents the number of integer bits and F is the number of fractional bits (e.g (1.15) notation) [EB99]. In addition, specialized hardware may be available to handle these formats within DSPs [FG00]. Most importantly, a designer should be aware of what precision he/she is working with so that when computing a result, the proper bits can be passed into the subsequent hardware element.

4.11 Summary

Carry-save array and tree multipliers are the two types of multipliers presented in this chapter. Both implementations have their trade-offs in terms of area and delay. Higher radix multipliers show an increase in performance over traditional multipliers at the expense of complexity. Implementations that utilize SD notation are also useful since they can perform carry-free addition. One implementation enables the final-carry propagate adder to be removed and are useful for recursive filters [EL90]. The techniques presented in this chapter have also been utilized within cryptographic systems as well [Tak92]. A nice review of multiplication methods can be found in [EL03].

```
module booth4x4 (P, A, X);

    input [3:0]  A, X;

    output [6:0] P;

    assign GND = 1'b0;
    // cas(cout, sout, U, R, Q, D, H, a, b, cin);
    cas m01(t00, s00, u00, r00, q00, r01, q01, A[0], GND, GND);
    cas m02(t01, s01, u01, r01, q01, r02, q02, A[1], GND, t00);
    cas m03(t02, s02, u02, r02, q02, r03, q03, A[2], GND, t01);
    cas m04(t03, s03, u03, r03, q03, r04, q04, A[3], GND, t02);
    cas m05(t10, s10, u10, r10, q10, r11, q11, u00, GND, GND);
    cas m06(t11, s11, u11, r11, q11, r12, q12, u01, s00, t10);
    cas m07(t12, s12, u12, r12, q12, r13, q13, u02, s01, t11);
    cas m08(t13, s13, u13, r13, q13, r14, q14, u03, s02, t12);
    cas m09(t14, s14, u14, r14, q14, r15, q15, u03, s03, t13);
    cas m10(t20, s20, u20, r20, q20, r21, q21, u10, GND, GND);
    cas m11(t21, s21, u21, r21, q21, r22, q22, u11, s10, t20);
    cas m12(t22, s22, u22, r22, q22, r23, q23, u12, s11, t21);
    cas m13(t23, s23, u23, r23, q23, r24, q24, u13, s12, t22);
    cas m14(t24, s24, u24, r24, q24, r25, q25, u14, s13, t23);
    cas m15(t25, s25, u25, r25, q25, r26, q26, u14, s14, t24);
    cas m16(t30, P[0], u30, r30, q30, r31, q31, u20, GND, GND);
    cas m17(t31, P[1], u31, r31, q31, r32, q32, u21, s20, t30);
    cas m18(t32, P[2], u32, r32, q32, r33, q33, u22, s21, t31);
    cas m19(t33, P[3], u33, r33, q33, r34, q34, u23, s22, t32);
    cas m20(t34, P[4], u34, r34, q34, r35, q35, u24, s23, t33);
    cas m21(t35, P[5], u35, r35, q35, r36, q36, u25, s24, t34);
    cas m22(t36, P[6], u36, r36, q36, r37, q37, u25, s25, t35);

    // Booth decoding
    ctrl c1(q37, r37, X[0], GND);
    ctrl c2(q26, r26, X[1], X[0]);
    ctrl c3(q15, r15, X[2], X[1]);
    ctrl c4(q04, r04, X[3], X[2]);

endmodule // booth4x4
```

Figure 4.28.   4-bit by 4-bit Signed Booth Carry-Save Array Multiplier.

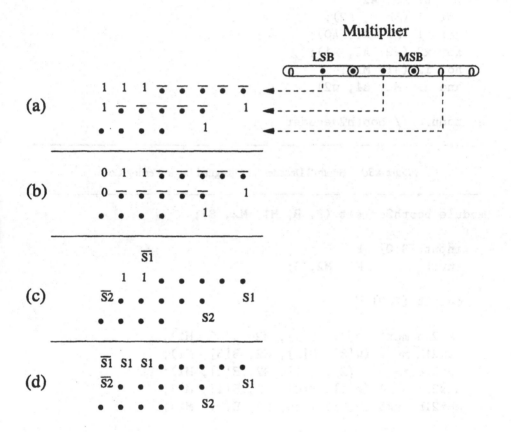

Figure 4.29. 4-bit by 4-bit Unsigned Radix-4 modified Booth Multiplier.

```
module booth2decoder (M1, M2, Sbar, S, A2, A1, A0);

    input A2, A1, A0;

    output M1, M2, Sbar, S;

    buf b1 (S, A2);
    not i1 (Sbar, A2);
    xor x1 (M1, A1, A0);
    xor x2 (w2, A2, A1);
    not i2 (s1, M1);
    and a1 (M2, s1, w2);

endmodule // booth2decoder
```

Figure 4.30. Booth-2 Decoder for Unsigned Logic Verilog Code.

```
module booth2select (Z, B, M1, M2, S);

    input [3:0]  B;
    input        M1, M2, S;

    output [4:0] Z;

    mux21h mux1 (w[4], B[3], M2, 1'b0, M1);
    mux21h mux2 (w[3], B[2], M2, B[3], M1);
    mux21h mux3 (w[2], B[1], M2, B[2], M1);
    mux21h mux4 (w[1], B[0], M2, B[1], M1);
    mux21h mux5 (w[0], 1'b0, M2, B[0], M1);

    xor x1 (Z[4], w[4], S);
    xor x2 (Z[3], w[3], S);
    xor x3 (Z[2], w[2], S);
    xor x4 (Z[1], w[1], S);
    xor x5 (Z[0], w[0], S);

endmodule // booth2select
```

Figure 4.31. Booth-2 Selector for Unsigned Logic Verilog Code.

```
module booth4 (Z, A, B);

    input  [3:0] A, B;

    output [7:0] Z;

    booth2decoder bdec1 (R1_1, R1_2, S1bar, S1, B[1],
                         B[0], 1'b0);
    booth2decoder bdec2 (R2_1, R2_2, S2bar, S2, B[3],
                         B[2], B[1]);
    booth2decoder bdec3 (R3_1, R3_2, S3bar, S3, 1'b0,
                         1'b0, B[3]);

    booth2select bsel1 (row1, A, R1_1, R1_2, S1);
    booth2select bsel2 (row2, A, R2_1, R2_2, S2);
    booth2select bsel3 (row3, A, R3_1, R3_2, S3);

    ha ha1 (fc3, fs2, row2[0], S2);
    ha ha2 (fc4, fs3, row2[1], row1[3]);
    fa fa1 (fc5, fs4, row2[2], row1[4], row3[0]);
    fa fa2 (fc6, fs5, row2[3], S1, row3[1]);
    fa fa3 (fc7, fs6, row2[4], S1, row3[2]);
    fa fa4 (fc8, fs7, S1bar, S2bar, row3[3]);

    rca8 cpa1 (Z, Cout,
               {fs7, fs6, fs5, fs4,
                  fs3, fs2, row1[1], row1[0]},
               {fc7, fc6, fc5, fc4,
                  fc3, row1[2], 1'b0, S1},
        1'b0);

endmodule // booth4
```

Figure 4.32.   Unsigned Radix-4 Multiplier Verilog Code.

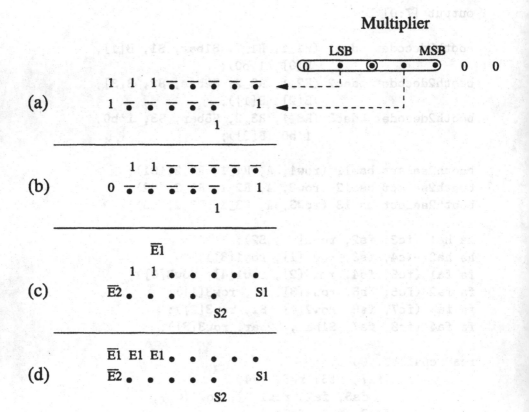

Figure 4.33.   4-bit by 4-bit Signed Radix-4 Multiplier.

```
module booth2select (Z, E, Ebar, B, M1, M2, S);

   input [3:0]  B;
   input        M1, M2, S;

   output [4:0] Z;
   output  E, Ebar;

   mux21h mux1 (w[4], B[3], M2, B[3], M1);
   mux21h mux2 (w[3], B[2], M2, B[3], M1);
   mux21h mux3 (w[2], B[1], M2, B[2], M1);
   mux21h mux4 (w[1], B[0], M2, B[1], M1);
   mux21h mux5 (w[0], 1'b0, M2, B[0], M1);

   xor x1 (Z[4], a1, w[4], S);
   xor x2 (Z[3], a2, w[3], S);
   xor x3 (Z[2], a3, w[2], S);
   xor x4 (Z[1], a4, w[1], S);
   xor x5 (Z[0], a5, w[0], S);
   xn x6 (Ebar, E, a6, S, B[3]);

endmodule // booth2select
```

Figure 4.34.   Booth-2 Selector for Signed Numbers Verilog Code.

```verilog
module booth4 (Z, A, B);

   input  [3:0] A, B;
   output [7:0] Z;

   wire [4:0]  row1, row2, row3;

   booth2decoder bdec1 (R1_1, R1_2, S1bar, S1, B[1],
                        B[0], 1'b0);
   booth2decoder bdec2 (R2_1, R2_2, S2bar, S2, B[3],
                        B[2], B[1]);

   booth2select bsel1 (row1, E1, E1bar, A, R1_1,
                        R1_2, S1);
   booth2select bsel2 (row2, E2, E2bar, A, R2_1,
                        R2_2, S2);

   ha ha1 (fc3, fs2, row2[0], S2);
   ha ha2 (fc4, fs3, row2[1], row1[3]);
   ha ha3 (fc5, fs4, row2[2], row1[4]);
   ha ha4 (fc6, fs5, row2[3], E1bar);
   ha ha5 (fc7, fs6, row2[4], E1bar);
   ha ha6 (fc8, fs7, E1, E2);

   rca8 cpa1 (Z, Cout,
      {fs7, fs6, fs5, fs4, fs3, fs2, row1[1], row1[0]},
      {fc7, fc6, fc5, fc4, fc3, row1[2], 1'b0, S1},
      1'b0);

endmodule // booth4
```

Figure 4.35.   Signed Radix-4 Multiplier Verilog Code.

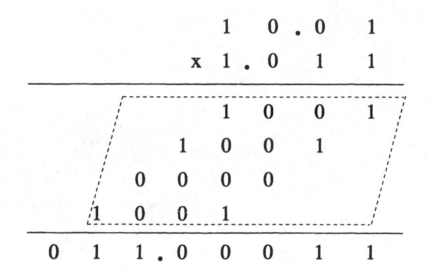

Figure 4.36.   4-bit by 4-bit Fractional Multiplication Matrix.

Chapter 5

DIVISION USING RECURRENCE

This chapter discusses implementations for division. There are actually many different variations on division including digit recurrence, multiplicative-based, and approximation techniques. This chapter deals with the class of division algorithms that are digit recurrence. Multiplicative-based division algorithms are explored in Chapter 7. For digit recurrence methods, the quotient is obtained one iteration at a time. In addition, the use of different radices are utilized to increase the throughput of the device. Although digit recurrence implementations are explored for division, the ideas presented in this chapter can also be utilized for square root, reciprocal square root, and online algorithms [EL94].

Similar to multiplication, the implementations presented here can be computed serially or in parallel. Although division is one of the most interesting algorithms that can be implemented in digital hardware, many designs have yet to match the speed of addition and multiplication units [OF97]. This occurs because there are many variations on a particular implementation including radix, circuit choices for quotient selection, and given internal precision for a design. Therefore, this chapter attempts to present several designs, however, it is important to realize that there many variations on the division implementations that make division quite exciting. With the designs in this chapter, a designer can hopefully get exposed to some of the areas in digit recurrence division in hopes of exploring more detailed designs.

As in multiplication, this chapter presents implementations of given algorithms for digit recurrence. There are many design variations on circuit implementations especially for quotient selection. Therefore, designs are presented in this chapter are presented in normalized fractional arithmetic (i.e. $1/2 \leq d \leq 1$). Variations on these designs can also be adapted for integer and floating-point arithmetic as well. Most of the designs, as well as square root, are discussed in [EL94] which is an excellent reference for many of these designs.

5.1   Digit Recurrence

Digit recurrence algorithms consist of n iterations where each iteration produces one digit of the quotient. However, digit recurrence algorithms do have preprocessing and postprocessing steps that are important to making sure that the dividend, divisor, quotient, and remainder are presented correctly.

Division starts by taking two components, the dividend, x, and the divisor, d, and computing the quotient, q, and its remainder, r. The main equation for division is recursive as follows:

$$x = q \cdot d + r \ni r < d \qquad (5.1)$$

The quotient digit is one of the most interesting elements to the division process. In particular, quotient digits are most often implemented as a redundant digit set. The reason for this is that it simplifies the quotient digit selection. Unfortunately, as the radix increases, the complexity of the quotient selection also increases. Therefore, as designers one of the challenges for division is to make conscientious decisions on an algorithmic implementation and its circuit implications. This trade off is probably magnified for division because of the variations in which division can be implemented.

Digit recurrence implementations of division work iteratively utilizing the following equation where w_i is the partial remainder for iteration i, d is the divisor, r is the radix, and q_i is the quotient digit for iteration i:

$$w_{i+1} = r \cdot w_i - q_{i+1} \cdot d$$

The dividend is inserted into the recurrence relationship by setting $w_0 = x$. The quotient selection function is chosen based on comparisons between the divisor and shifted partial remainder:

$$q_{i+1} = QST(r \cdot w_i, d)$$

where QST is the Quotient Selection Table . The QST can be implemented differently including ROM tables, PLAs, and combinational logic [SL95]. In order to save space and make the implementation straight forward, the designs presented in this chapter utilizes combinational elements. In addition, there are also methods for exploiting symmetry within the QST to reduce the hardware requirements [OF98].

As stated previously, the most challenging steps in the division procedure is the comparison between the divisor and the remainder to determine the quotient bit. If this is done by subtracting d from w_i, one has to be careful if the result is negative. If so, a correction operation occurs restoring the remainder to the previous iteration. This method is called restoring division. Non-restoring division is an alternative for sequential division by having specific logic for not correcting the quotient. This is achieved by allowing a correction factor within the algorithm. Unfortunately, because non-restoring division requires a correction factor, there may be some post-processing that is required if the final remainder is negative. Consequently, it is necessary to have a correction step that adjusts the quotient as follows where m is the final iteration of the recurrence relation and r^{-n} is an ulp:

$$q = \begin{cases} q_m & \text{if } w_i \geq 0 \\ q_m - r^{-n} & \text{if } w_i < 0 \end{cases} \tag{5.2}$$

Therefore, the process of division by recurrence can improve upon general division algorithms by taking advantage of the following elements [EL03]

1. Radix decreases the number of iterations assuming $r = 2^k$ by $\lfloor log_2(2^k) \rfloor$

2. Redundancy within the quotient digit set reduces and simplifies the QST

3. Partial remainder can be implemented using redundant notation which simplifies the computation of the partial remainder using a carry-free adder

The designs presented in this chapter were chosen in hopes of illustrating each of these benefits and illustrating the trade-offs for these choices. Since the designs in this exploration iteratively decide the quotient, efficient control logic is required to help the datapath perform correctly. Since control logic is usually completed last in the design phase as described in Chapter 1, having a table of control lines and times needed to implement this logic makes the task simple and efficient.

5.2 Quotient Digit Selection

The algorithm for division is challenging because the implementations for the quotient digit selection vary from design to design. The basic idea of the QST is to choose the value of the quotient digit, q_{i+1}, based on a comparison between the shifted partial remainder and the divisor. A symmetric SD digit set is utilized where the range of digits is

$$q_i \in \{\overline{\alpha}, \overline{\alpha} + 1, \ldots \overline{1}, 0, 1, \ldots, \alpha - 1, \alpha\}$$

with the measure of redundancy, ρ and radix r, defined as follows:

$$\rho = \frac{\alpha}{r - 1} \ni \frac{1}{2} < \rho \leq 1$$

Although choosing the right function for a QST is complex, it can easily be formulated into two conditions called *containment* and *continuity*. The containment condition determines the selection interval for each quotient digit,

q_{i+1}[EL94]. On the other hand, continuity condition details the range which the quotient digit is selected [EL94].

5.2.1   Containment Condition

Since the equation for the recurrence involves subtractions and shifts, it is important that the quotient digit selection becomes difficult. For example, if a user decided to divide $400 \div 5$ in radix 10 and inadvertently chooses q_1 to be 2. This will violate the bounds available for the next quotient digit making the computation cumbersome. In other words, a quotient digit will need to be computed for the partial remainder of 300.

The containment condition sets up the selection intervals necessary for computing the subsequent quotient digit. For a given quotient digit, q_{i+1} can be chosen to be k. Therefore, an interval of allowable partial remainders. These regions are defined by the interval $[L_k, U_k]$ such that L is the lower value and U is the upper value of the partial remainder, $r \cdot w_i$, so that the subsequent shifted partial remainder is bounded. In other words, the interval is chosen based on the range of redundancy:

$$U_k = (k + \rho) \cdot d$$
$$L_k = (k - \rho) \cdot d$$

Sometimes, this can visualized by examining a graph of the subsequent partial remainder, w_{i+1}, versus the shifted partial remainder, $r \cdot w_i$. This visualization is represented in Robertson's diagram as shown in Figure 5.1 [Rob58]. Robertson's diagram plots the recurrence relationship for a given quotient digit, q_{i+1}, assuming the user is varying the shifted partial remainder and plotting or computing the subsequent partial remainder. The axis of Robertson's diagram is bounded by $axis([-r\rho \cdot d \; ; \; r\rho \cdot d \; ; \; -\rho \cdot d \; ; \; \rho \cdot d].')$ where the *axis* function defines the range of the function such that the argument is defined as (*[xmin ; xmax ; ymin ; ymax]*). Interestingly, the redundancy introduced by using a SD digit set imposes an overlap between quotient digits. For example, in Figure 5.1 there is an overlap between $q_{i+1} = k - 1$ and $q_{i+1} = k$. This overlap will be useful in defining the continuity equation.

5.2.2   Continuity Condition

Since the containment condition defines the range of the subsequent partial remainder, choosing the correct quotient digit from this region. This is the job of the continuity condition. To satisfy the containment condition, the minimum value of the x axis of the Robertson's diagram is chosen such that $q_{i+1} = k$ is our quotient digit [EL94]. This can be defined as the following inequality where s_k is our minimum value that a user chooses before an

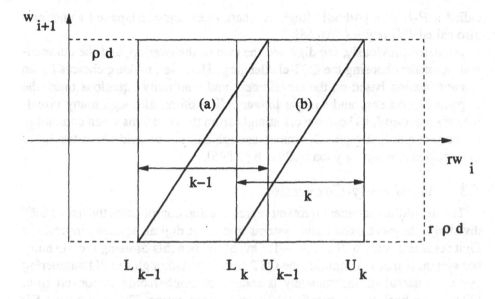

Figure 5.1. Robertson Diagram for (a) $q_{i+1} = k - 1$ and (b) $q_{i+1} = k$.

implementation is devised

$$L_k \leq s_k \leq U_k$$

Unfortunately, because the overlap that occurred in the containment condition, a quotient digit may be chosen from either minimum value. For example, in Figure 5.1, an overlap exists between L_k and U_{k-1} such that s_k can either be $k - 1$ or k. Since the containment equations are defined, it is easy to measure this overlap as

$$U_{k-1} - L_k = (k - 1 + \rho \cdot d) - (k - \rho \cdot d) = (2 \cdot \rho - 1) \cdot d$$

The simplest selection function is to make s_k constant and do a comparison on the constant. Thus, many implementations for QSTs resort to ROM tables or PLA elements. The constants should satisfy the following equation [EL94]:

$$max(L_k) \leq m_k \leq min(U_{k-1}) + ulp$$

It should be obvious that a given digit set does not have to an overlap. However, an overlap means that a given implementation may be different for each designer since an overlap region can be large. On the other hand, the main reason for having a redundant quotient digit set is to provide a overlap to simplify the QST. Some have suggested plotting the shifted partial remainder, $r \cdot w_i$ versus the divisor to visualize the overlap regions easier. This kind of plot is called a P-D plot [Atk68]. Regions where there are overlaps are sometimes also called PD regions [Atk68].

As stated previously, the digit set, the size of the overlap, and the comparison can make choosing the QST challenging. However, making choices for an implementation based on the containment and continuity equations make the implementation easy and straight forward. Therefore, although many visualizations are useful, its best to work straight from the equations when computing the correct quotient digit. Of course, enough can not be said about testing of these regions through a good testbench [CT95].

5.3   On-the-Fly-Conversion

The use of the redundant quotient representation complicates the use of SRT division.   In most fixed radix systems that most digital devices employ, the digit set is restricted to $0, \ldots, r - 1$. One of the benefits of using the SD number system is that is simplifies the QST [Atk68]. Although the SD numbering system is useful, it unfortunately is somewhat cumbersome to convert from SD notation back to a conventional binary representation. To convert from SD notation to conventional binary representation involves the use of a CPA.

Fortunately, division utilizing recurrence equations computes the quotient with the Most Significant Digit First (MSDF). Arithmetic performed in this manner is sometimes referred to as online arithmetic [EL92a], [EL94]. Since the quotient is computed as a fraction the quotient is computed as

$$q_i = \sum_{m=1}^{i} q_m \cdot r^{-m}$$

Therefore, using the correction factor and plugging it into the equation above results in the following form [EL92b]:

$$q_{i+1} = \begin{cases} q_i + q_{i+1} \cdot r^{-(i+1)} & q_{i+1} \geq 0 \\ q_i - r^{-j} + (r- \mid q_{i+1} \mid) \cdot r^{-(i+1)} & q_{i+1} < 0 \end{cases}$$

The latter equation is formed since the quotient for that iteration is negative. Therefore, a subtraction is required for the conversion. If we substitute a variable for the correction factor, qm_i, the equation above is presented more efficiently as:

$$q_{i+1} = \begin{cases} q_i + q_{i+1} \cdot r^{-(i+1)} & q_{i+1} \geq 0 \\ qm_i + (r- \mid q_{i+1} \mid) \cdot r^{-(i+1)} & q_{i+1} < 0 \end{cases}$$

With simple manipulation, we can also convert the equation above into an equation for qm_i such that $qm_i = q_i - r^{-n}$. In other words, if the final remainder is negative, subtraction of an ulp from the quotient is performed to adjust the correction factor. Therefore, qm_i is computed as follows:

$$qm_{i+1} = \begin{cases} q_i + (q_{i+1} - 1) \cdot r^{-(i+1)} & q_{i+1} > 0 \\ qm_i + ((r-1) - \mid q_{i+1} \mid) \cdot r^{-(i+1)} & q_{i+1} \leq 0 \end{cases}$$

Fortunately, there is an easy algorithm to convert back to conventional representation from SD notation for on-line algorithms. It is called on-the-fly conversion [EL92b]. The basic idea behind on-the-fly conversion is to produce the conversion as the digits of the quotient are produced by performing a concatenation instead of any carries or borrows within a carry-propagate adder. One element keeps track of the normal *quotient*, whereas, another element keeps track of the *quotient − ulp*. This technique is very similar to the carry-select logic in the carry-select adder from Chapter 3.

Since on-the-fly conversion involves concatenations, the MSDF enables the appropriate quotient digit to be converted by simple combinational logic and shifting as opposed to utilizing a CPA. The algorithm can be summarized as follows in terms of concatenations.

$$q_{i+1} = \begin{cases} \{q_i, q_{i+1}\} & \text{if } q_{i+1} \geq 0 \\ \{qm_i, (r - \mid q_{i+1} \mid\} & \text{if } q_{i+1} < 0 \end{cases}$$

and

$$qm_{i+1} = \begin{cases} \{q_i, q_{i+1} - 1\} & \text{if } q_{i+1} > 0 \\ \{qm_i, ((r-1) - \mid q_{i+1} \mid\} & \text{if } q_{i+1} \leq 0 \end{cases}$$

In order to implement on-the-fly conversion, it requires two registers to contain q_i and qm_i. These registers are shifted one digit left with insertion into the least-significant digit, depending on the value of q_{i+1} In other words, depending on the what the subsequent quotient digit, the register either chooses q or qm and concatenates the current converted quotient digit into the least-significant digit. Figure 5.2 shows the basic structure for radix 2 on the fly conversion for 8 bits (i.e. $n = 8$). Two multiplexors are utilized to select either q or qm and combinatorial logic is used to select q_{in} and qm_{in}. In order to handle shifting after every cycle, the output of the multiplexors are shifted by one (multiplied by 2) and either q_{in} or qm_{in} are inserted into the least significant bit during each load. The final multiplexor choose the correct quotient once the final remainder is known. If the sign of the final remainder is 1, it will choose qm since this register contains the proper corrected quotient. Finally, q^* and qm^* are shown in Figure 5.2 to designate q_{i+1} and qm_{i+1}, respectively.

For radix 2 on-the-fly conversion, the registers are updated according to the values in Table 5.2. The values in this table are computed utilizing the equations above for qm_i and q_i and $r = 2$. For this example, the quotient digit

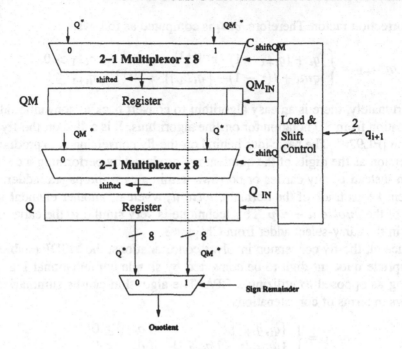

Figure 5.2. Radix 2 On-the-Fly-Conversion Hardware assuming $n = 8$.

utilized is $\{\bar{1}, 0, 1\}$. The values of C_{shiftq} and $C_{shiftqm}$ are used to control the multiplexors. The value of q_{in} and qm_{in} is the concatenation element input into the register. The quotient is utilized as input to compute C_{shiftq}, $C_{shiftqm}$, q_{in}, and qm_{in}. In order to simplify the logic, the quotient utilizes one-hot logic encoding as shown in Table 5.1 since this form of encoding introduces *don't cares*. This form of encoding for digit recurrence division is popular and similar to the designs found in Chapter 4. Therefore, the equations

quotient	q^+	q^-
$\bar{1}$	0	1
0	0	0
1	1	0

Table 5.1. Quotient Bit Encoding.

for computing C_{shiftq}, $C_{shiftqm}$, q_{in}, and qm_{in} for radix 2 and the encoding shown in Table 5.1 are straight forward and can be computed using simple

Boolean two-level simplification as shown below.

$$
\begin{aligned}
C_{shiftq} &= q_{i+1}[0] \\
C_{shiftqm} &= q_{i+1}[1] \\
q_{in} &= q_{i+1}[0] + q_{i+1}[1] \\
qm_{in} &= \overline{q_{i+1}[0] + q_{i+1}[1]}
\end{aligned}
$$

For example, suppose conversion is required for the following SD number 1101$\bar{1}$00 to a conventional representation using on-the-fly conversion. Table 5.3 shows how on-the-fly-conversion works is updated according to Table 5.2. At step $i = 0$, the values for both registers are reset which can be accomplished by using a flip-flop that has reset capabilities. In addition, since division is an online algorithm, on-the-fly conversion works from the most-significant bit to the least-significant bit. The last value in the register is the final converted value assuming a fractional number for q_i and qm_i which is 0.78125 and 0.7734375, respectively. It should be obvious that both of these elements are one *ulp* from each other (i.e. an *ulp* in this case is 2^{-7} or 0.0078125) and 0.78125 is the conventional representation of 1101$\bar{1}$00.

The Verilog code for the radix 2 on-the-fly conversion is shown in Figure 5.3. The instantiation for $ls_control$ implements the Boolean logic for the computing C_{shiftq}, $C_{shiftqm}$, q_{in}, and qm_{in} although this instantiation could probably be avoided by utilizing bitwise operators. The variables $Qstar$ and $QMstar$ are utilized to make the subsequent iteration of the conversion easy to debug. Since the register inside the conversion logic may potentially clash in terms of timing with other parts of the datapath, careful timing is involved. For flip-flop based registered, multi-phase clocking is usually best to handle the conversion and quotient selection. On the other hand, retiming of the recurrence by using advanced timing methodologies can lead to lower a power dissipation [NL99].

q_{i+1}	q_{in}	C_{shiftq}	q_{i+1}	qm_{in}	$C_{shiftqm}$	qm_{i+1}
1	1	1	$\{q_i, 1\}$	0	0	$\{q_i, 0\}$
0	0	1	$\{q_i, 0\}$	1	1	$\{qm_i, 1\}$
$\bar{1}$	1	0	$\{qm_i, 1\}$	0	1	$\{qm_i, 0\}$

Table 5.2. Radix-2 on-the-fly-conversion.

i	q_i	q	qm
0	0	0	0
1	1	0.1	0.0
2	1	0.11	0.10
3	0	0.110	0.101
4	1	0.1101	0.1100
5	$\bar{1}$	0.11001	0.11000
6	0	0.110010	0.110001
7	0	0.1100100	0.1100011

Table 5.3.   Example On-The-Fly Conversion.

```
module conversion (Q, quot, SignRemainder, Clk, Load, Reset);

    input   [1:0] quot;
    input         SignRemainder, Clk, Load, Reset;

    output [7:0] Q;

    ls_control ls1(Qin, QMin, CshiftQ, CshiftQM, quot);
    mux21x8  m1(M1Q, Qstar, QMstar, CshiftQM);
    register8 r1(R1Q, {M1Q[6:0], QMin}, Clk, Load, Reset);
    mux21x8  m2(M2Q, QMstar, Qstar, CshiftQ);
    register8 r2(R2Q, {M2Q[6:0], Qin}, Clk, Load, Reset);
    mux21x8  m3(Q, R2Q, R1Q, SignRemainder);
    assign Qstar = R2Q;
    assign QMstar = R1Q;

endmodule // conversion
```

Figure 5.3.   Radix 2 On-the-Fly Conversion Verilog Code.

5.4   Radix 2 Division

The radix-2 or binary division algorithm is quite easy to implement. The main elements that are needed are an adder to add or subtract the partial remainder, multiplexors, registers, and some additional combinatorial devices. Figure 5.4 shows the basic block diagram for the design. The algorithm presented here is an extension of non-restoring division with a quotient digit set of $\{\bar{1}, 0, 1\}$. It utilizes an adder to add the partial remainder in Non-redundant

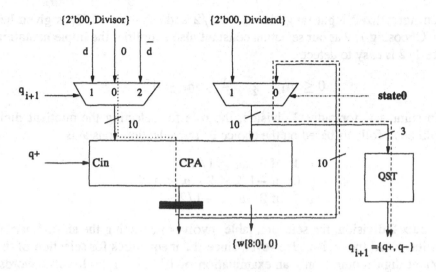

Figure 5.4. Radix 2 Division.

form. This type of algorithm is sometimes called SRT division [Rob58], [Toc58]. The implementation stems from the basic recurrence relationship

$$w_{i+1} = 2 \cdot w_i - q_{i+1} \cdot d$$

SRT division was named after Sweeney, Robertson, and Tocher, each of whom developed the idea independently from each other [Rob58], [Toc58]. The main goal is to speed up division by allowing a 0 as a quotient digit. This eliminates the need for a subtraction or addition when the value 0 is selected.

To start implementing the QST, the containment condition must first be computed. Using the equations for containment and $\rho = 1$, the following condition exists

$$
\begin{aligned}
L_1 &= 0 & U_1 &= 2 \cdot d \\
L_0 &= -d & U_0 &= d \\
L_{-1} &= -2 \cdot d & U_{-1} &= 0
\end{aligned}
$$

Using these elements within the continuity condition presents the following inequalities

$$
\begin{aligned}
0 &\le s_1 \le d \\
-d &\le s_0 \le 0
\end{aligned}
$$

One possible implementation is to choose constants, m_k, from the region defined by s_k. One set of constants that satisfy this requirement, assuming the given normalized input range is $m_1 = 1/2$ and $m_{-1} = -1/2$, is given below. Choosing $1/2$ as our selection constant also simplifies the implementation since $1/2$ is easy to detect.

$$0 \leq m_1 \leq \tfrac{1}{2} \qquad \tfrac{-1}{2} \leq m_0 \leq 0$$

In summary, for radix 2 division, the rule for selecting the quotient digit would be as follows based on the choice of our selection constants:

$$q_i = \begin{cases} 1 & \text{if } 2 \cdot w_i \geq 1/2 \\ 0 & \text{if } -1/2 \leq 2 \cdot w_i < 1/2 \\ \bar{1} & \text{if } 2 \cdot w_i < -1/2 \end{cases}$$

For radix 2 division, the selection table involves inspecting the shifted partial remainder and the divisor. However, since the inequalities for selection of the quotient digit requires only an examination of $1/2$ or $-1/2$. In other words, only the most significant fractional bit requires examination. To examine $-1/2$ a sign bit is required making only two bits necessary for examination instead of a full-length comparison. However, since the maximum range is $\rho * r \cdot d$ or $2 \cdot d$, an integer bit is required to make sure the partial remainder fall in range. Therefore, 3 bits of $2 \cdot r_i$ must be examined every iteration.

Table 5.4 tabulates the bits to compare for the proper quotient digit. As in the example in on-the-fly conversion, one-hot encoding is utilized for the quotient digit. The bit q_+ signifies a positive bit or 1, whereas, q_- is a negative bit or $\bar{1}$. Using simple Boolean simplificaiton, the quotient digit selection is as follows:

```
assign q[1] = (!Sign&Int) | (!Sign&f0);
assign q[0] = (Sign&!Int) | (Sign&!f0);
```

where $Sign$ is the sign bit, Int is the integer bit, and $f0$ is the first fractional bit. For an 8-bit operand, $Sign$, Int, and $f0$ are $d[7]$, $d[6]$, and $d[5]$, respectively. Bitwise operators are utilized to build the Verilog code.

As stated previously, the block diagram is shown in Figure 5.4. A CPA is utilized to add or subtract the shifted partial remainder for the computation of the subsequent partial remainder. The worst-case delay is illustrated by the dotted line. Since a CPA consume a large amount of delay especially for larger operand sizes, carry-free addition schemes are often opted for. However, utilized carry-free adders introduces error into the partial remainder computation, thereby, increasing the complexity of the QST. The control signal $state0$ is utilized to initialize the registers in the first iteration. In order to get the first iteration bounded appropriately, the dividend also needs to be scaled by $1/2$. This can be performed without a correction, because it means that the quotient will be still be correct yet shifted.

sign	int	f0	Result	Quotient	$q+$	$q-$
0	0	0	$< 1/2$	0	0	0
0	0	1	$\geq 1/2$	1	1	0
0	1	0	$\geq 1/2$	1	1	0
0	1	1	$\geq 1/2$	1	1	0
1	0	0	$< -1/2$	-1	0	1
1	0	1	$< -1/2$	-1	0	1
1	1	0	$< -1/2$	-1	0	1
1	1	1	$\geq -1/2$	0	0	0

Table 5.4. Quotient Digit Selection.

The Verilog code is shown in Figure 5.5. It is assumed that the input operands are not negative. Division by recurrence can be extended to include negative operands in two's complement by examining the quotient and the divisor. Although the input is 8 bits, internally the recurrence computes with 10 bits (i.e. one additional integer bit and a sign bit). Utilizing internal precision that is larger than the input or output operands is typical in many arithmetic datapaths. The on-the-fly conversion and other ancillary logic such as zero detection are not shown, yet can be added easily. Zero detection is needed for appropriate rounding [IEE85]. Since the logic needs selection logic for d, $\overline{d}$ and 0, a 3-1 multiplexor is required. This multiplexor can be extended from the version in Chapter 3 as shown in Figure 5.6. Other multiplexors, such as 4-1 can be built this way as well. The 10 bit register, shown by the filled rectangle, is 10 instantiations of the *dff* module found in Chapter 2 and implemented similar to the 2-bit multiplexor found in Chapter 3. If $q_{i+1} = q[1] = 1$, the recurrence relation indicates a subtraction is necessary ($w_{i+1} = 2 \cdot w_i - d$, therefore, the 3-1 multiplexor selects the one's complement of d. In order to complete the subtraction, the partial remainder must be in two's complement form. Therefore, c_{in} is asserted appropriately into the CPA for this particular quotient digit. Moreover, the variable *zero* is utilized to decode the 0 quotient digit set, because the one-hot encoding.

5.5 Radix 4 Division with $\alpha = 2$ and Non-redundant Residual

Radix 4 is another possible implementation [Rob58], [BH01]. Since $r = 4$, the value of α can be 2 or 3. Choosing $a = 2$ makes the generation of $q_i \cdot d$ easier. Several efficient implementations have been proposed for $\alpha = 3$ as well [SP94]. Therefore, the quotient digit utilized will be $\{\overline{2}, \overline{1}, 0, 1, 2\}$. To implement this datapath, it is necessary to first formulate the containment and

```
module divide2 (w, q, D, X, state0, Clk);

    input [7:0] D, X;
    input Clk, state0;

    output [9:0] w;
    output [1:0] q;

    inv10 i1 (OnesCompD, {2'b00,D});
    mux31x10 m3 (Mux3Out, 10'b0000000000, {2'b00, D},
                OnesCompD, q);
    mux21x10 m2 (Mux2Out, {Sum[8:0], 1'b0}, {2'b00, X},
                state0);
    reg10 r1 (w, Mux2Out, Clk);
    // q = {+1, -1} else q = 0
    qst qd1 (q, w[9], w[8], w[7]);
    rca10 d1 (Sum, w, Mux3Out, q[1]);

endmodule // divide2
```

Figure 5.5.   Radix 2 Division Verilog Code.

continuity equations based on $\rho = 2/(4 - 1) = 2/3$. Since $\rho = 2/3$, the containment condition results in the following selection intervals:

$$L_k = \left(K - \tfrac{2}{3}\right) \cdot d \quad U_k = \left(k + \tfrac{2}{3}\right) \cdot d$$

For this implementation, we can no longer implement the QST based only on the shifted partial remainder. The QST will be based on the δ most-significant bits of d, however, since the implementations presented here are for normalize fractions $1/2 \le d \le 1$, only $\delta - 1$ bits are required since the most significant fractional bit is a 1. Therefore, the quotient digit selection can be chosen based on [EL94]:

$$m_k(i) \ge A_k(i) \cdot 2^{-c}$$

where c represents the total fractional bits necessary to support the continuity equation with m_k. $A_k(i)$ is an integer that selects the proper boundary for the quotient digit selection from the given bounds on the partial remainder. However, since the QST is based on the size of the radix and selection of α, the quotient digit selection is no longer a single constant for a given shifted

```
module mux31 (Z, A, B, C, S);

   input [1:0] S;
   input       A, B, C;

   output Z;

   not not1 (s0bar, s[0]);
   not not2 (s1bar, s[1]);
   nand na1 (w1, s0bar, s1bar, A);
   nand na2 (w2, B, s[0], s1bar);
   nand na3 (w3, C, s0bar, s[1]);
   nand na4 (Z, w1, w2, w3);

endmodule // mux31
```

Figure 5.6. 3-1 Multiplexor Verilog Code.

partial remainder and divisor. It usually resembles a function that resembles a staircase. The value of δ can be found from the following equation [EL94]:

$$2^{-\delta} = \frac{2 \cdot \rho - 1}{2 \cdot \rho \cdot (r - 2)}$$

and, the value of c can be obtained from the following inequality

$$L_k(d_{i+1}) \leq m_k(i) < U_{k-1}(d_i) \quad \text{for } k > 0$$
$$L_k(d_i) \leq m_k(i) < U_{k-1}(d_{i+1}) \quad \text{for } k \leq 0$$

As a datapath design, the overall idea is to minimize both c and δ so that the implementation of the QST is small since the QST can limit the implementation cost [OF98]. Visually, the QST now looks like Figure 5.7. The value of $log_2(r)$ comes from the fact that the bounds of Robertson's diagram is $r \cdot \rho \cdot d$.

For the radix 4 implementation, the bound on δ is

$$\delta \geq 3 \rightarrow 2^{-\delta} \geq \frac{1}{8}$$

However, as in radix 2 a careful choice of m_k is required. For example, having $m_k = 1/3$ is not a prudent choice because it would require a full precision comparison. For example, for radix 4 and $\delta = 3$ and inspection of the divisor range of $d = [4/8, 5/8]$, $L_2(5/8) = 4/3 \cdot 5/8 = 20/24$ and $U_1(4/8) =$

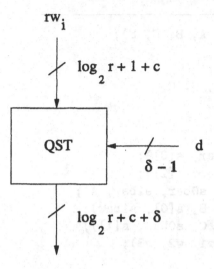

Figure 5.7. QST Block Diagram.

$5/3 \cdot 4/8 = 20/24$ which illustrates a potential problem with the comparison. Unfortunately, $20/24$ requires a full comparison for $d = 3$ bits. Moreover, the continuity relation indicates that $20/24 < 20/24$ which is not possible. Therefore, $\delta = 4$ is chosen to guarantee that all quotient digits can be selected appropriately.

5.5.1   Redundant Adder

This particular implementation is different than the radix 2 divider in that a carry-save adder is utilized to speed up the partial remainder computation. The use of a carry-save adder as opposed to a CPA introduce error into the computation, therefore, the containment and continuity equations must be modified. In this implementation a carry-save adder will be utilized to keep the residual in Non-redundant form, however, a SD adder could also be used.

The introduction of the carry save adder and the use of two's complement numbers produces error as opposed to a CPA. The error due to truncation, fortunately, is always positive with carry-save adders, however, there will be small amount of error introduced where t is the number of fractional bits:

$$0 \leq \epsilon \leq 2^{-t+1} - ulp$$

Therefore, the containment condition gets modified to:

$$\hat{L}_k = L_k$$
$$\hat{U}_k = U_k - 2^{-t+1} + ulp$$

Since the continuity condition relates the largest value for which it is still possible to choose $q_{i+1} = k - 1$, the upper and lower bound are modified for the carry save adder to:

$$\widehat{U}_{k-1} = U_{k-1} - 2^{-t}$$
$$\widehat{L}_k = L_k$$

Subsequently, the continuity condition produces new intervals for the selection of the quotient digits truncated t ɔ t bits:

$$\widehat{L}_k(d_{i+1}) \le m_k(i) < \widehat{U}_{k-1}(d_i) \quad \text{for } k > 0$$
$$\widehat{L}_k(d_i) \le m_k(i) < \widehat{U}_{k-1}(d_{i+1}) \quad \text{for } k \le 0$$

These new inequalities also produce new constraints on the divisor as well [EL94]. Therefore, the corresponding expression for t and δ is

$$\frac{2 \cdot \rho - 1}{2} - (a - \rho) \cdot 2^{-\delta} \ge 2^{-t}$$

Moreover, the range of the estimate is also modified since the new boundary on the Robertson's diagram shrinks. The new range is

$$-r \cdot \rho - 2^{-t} \le r \cdot \hat{y} \le r \cdot \rho - ulp$$

5.6 Radix 4 Division with $\alpha = 2$ and Carry-Save Adder

Using the new formulations for radix 4 division with $\alpha = 2$ and $\rho = 2/3$ adjusts the requirements for the number of bits for c and δ. Using the relationship for δ and t produces

$$\frac{1}{6} - \frac{4}{3} \cdot 2^{\delta} \ge 2^{-t}$$

Using $t = 4$ and $\delta = 4$ satisfies this inequality. The new containment and continuity equations produce a nice implementation for $c = 4$ as shown in Table 5.5 [EL94]. Therefore, the total number of bits that are examined for the shifted partial remainder is $4 + 2 + 1 = 7$ and the total number of bits for the divisor is 4. However, since the divisor is normalized, the leading most-significant bit does not need to be stored.

The implementation of the QST using Table 5.5 is done similar to the radix 2 implementation using combinational logic. However, in order to save space, the implementation uses a technique for reducing the logic of the QST. Since a carry-save adder introduces

$$0 \le \epsilon \le 2^{-t+1} - ulp$$

$[d_i, d_{i+1})$	$[8,9)$	$[9,10)$	$[10,11)$	$[11,12)$
$\hat{L}_2(d_{i+1}), \hat{U}_1(d_i)$	12, 12	14, 14	15, 15	16, 17
$m_2(i)$	12	14	15	16
$\hat{L}_1(d_{i+1}), \hat{U}_0(d_i)$	3, 4	4, 5	4, 5	4, 6
$m_1(i)$	4	4	4	4
$\hat{L}_0(d_{i+1}), \hat{U}_{\bar{1}}(d_i)$	−5, −4	−6, −5	−6, −6	−7, −5
$m_0(i)$	−4	−6	−6	−6
$\hat{L}_{\bar{1}}(d_{i+1}), \hat{U}_{\bar{2}}(d_i)$	−13, −13	−15, −15	−16, −16	−18, −17
$m_{\overline{-1}}(i)$	−13	−15	−16	−18
$[d_i, d_{i+1})$	$[12,13)$	$[13,14)$	$[14,15)$	$[15,16)$
$\hat{L}_2(d_{i+1}), \hat{U}_1(d_i)$	18, 19	19, 20	20, 22	22, 24
$m_2(i)$	18	20	20	24
$\hat{L}_1(d_{i+1}), \hat{U}_0(d_i)$	5, 7	5, 7	5, 8	6, 9
$m_1(i)$	6	6	8	8
$\hat{L}_0(d_{i+1}), \hat{U}_{\bar{1}}(d_i)$	−8, −6	−8, −6	−9, −6	−10, −7
$m_0(i)$	−8	−8	−8	−8
$\hat{L}_{\bar{1}}(d_{i+1}), \hat{U}_{\bar{2}}(d_i)$	−20, −19	−21, −20	−23, −21	−25, −23
$m_{\overline{-1}}(i)$	−20	−20	−23	−21

Table 5.5. Selection Intervals and Constants for Quotient Digit Selection shown as $d_{real} = d_{shown}/16$ (Values adapted from [EL94]).

of error and a carry-save output are input into the QST, we can mitigate this error by putting a small CPA to add the sum and carry parts as shown in Figure 5.8. This subsequently reduces the number of product terms from 5,672 to 45 making the implementation simple at the expense of adding additional logic. Similar to the radix 2 implementation, one-hot encoding is utilized as shown in Table 5.6. The Verilog code for the new complete QST is not shown but is easily created by inputting Table 5.5 into a Boolean minimizer, such as Espresso, and utilizing bitwise operators [SSL+92]. For example, the following code is the quotient bit obtained from Espresso for q^{2+} in Table 5.6.

```
assign q[3] = (!s[6]&s[5]) | (!d[2]&!s[6]&s[4]) |
              (!s[6]&s[4]&s[3]) |
              (!d[1]&!s[6]&s[4]&s[2]) |
              (!d[0]&!s[6]&s[4]&s[2]) |
              (!d[1]&!d[0]&!s[6]&s[4]&s[1]) |
              (!d[2]&!d[1]&!d[0]&!s[6]&s[3]&s[2]) |
              (!d[2]&!d[1]&!s[6]&s[3]&s[2]&s[1]) |
              (!d[2]&!d[0]&!s[6]&s[3]&s[2]&s[1]&s[0]);
```

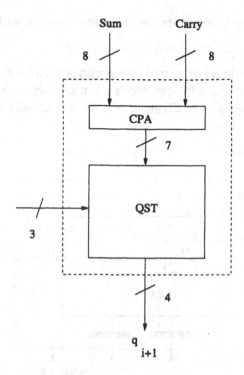

Figure 5.8. New Selection Function.

quotient	q^{2+}	q^+	q^-	q^{2-}
$\bar{2}$	0	0	0	1
$\bar{1}$	0	0	1	0
0	0	0	0	0
1	0	1	0	0
2	1	0	0	0

Table 5.6. Quotient Bit Encoding.

The block diagram of the implementation is shown in Figure 5.9. The dotted line indicates the worst-case delay. Two registers are now required to save the partial remainder in carry-save form. In addition, two 2-1 multiplexors are now required to initialize the carry-save registers. A 4-1 multiplexor choose the proper shifted divisor. Since the ulp can no longer be asserted into the carry-save adder (CSA) unless additional logic is utilized, the ulp is concatenated on the end of the carry-portion of the shifted partial remainder. Since the implementation now implements the following equation with

$$w_{i+1} = 4 \cdot w_i - q_{i+1} \cdot d$$

both the partial remainder carry and save portions must be shifted by two every iteration. Consequently, only four cycles are necessary to complete the recurrence without rounding. The dividend, similar to the radix 2 implementation, requires to be shifted by 4.

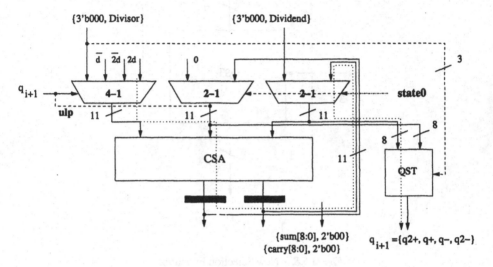

Figure 5.9. Radix 4 Division.

The Verilog code for the radix 4 divide unit is shown in Figure 5.10. Concatenation is utilized to get the divisor and dividend in the right format. As can be seen by the radix 4 implementation, the improvement in execution is achieved at the expense of complexity. As the radix increases, more logic is required to handle the complexity in decoding and encoding the QST. On the other hand, if the QST can sustain a low amount of complexity, the radix 4 implementation can be efficient.

5.7   Radix 16 Division with Two Radix 4 Overlapped Stages

Unfortunately, as the radix increases, so does the complexity of the QST. As the radix surpasses 8, it becomes obvious that other methods are required to obtain increased efficiency with a moderate amount of hardware complexity. One such modification is the use of overlapping stages [PZ95], [Tay85]. In this method, two stages are utilized to split the quotient digit into a high and low part. In using two radix 4 stages to create a radix 16 implementation $q_i = 4 \cdot q_H + q_L$ with a digit set of $\{\overline{2}, \overline{1}, 0, 1, 2\}$. Theoretically, this makes the resulting digit set $[-10, 10]$ or $\alpha = 10$.

```
module divide4 (quotient, Sum, Carry, X, D2, clk, state0);

    input [7:0]   X, D;
    input    clk, state0;

    output [3:0]   quotient;
    output [10:0] Sum, Carry;

    assign divi1 = {3'b000, D};
    assign divi2 = {2'b00, D, 1'b0};
    inv11 inv1 (divi1c, divi1);
    inv11 inv2 (divi2c, divi2);
    assign dive1 = {3'b000, X};

    mux21x11 mux1 (SumN, {Sum[8:0], 2'b00}, dive1, state0);
    mux21x11 mux2 (CarryN, {Carry[8:0], 2'b00},
                   11'b00000000000, state0);
    reg11 reg1 (SumN2, SumN, clk);
    reg11 reg2 (CarryN2, CarryN, clk);
    rca8 cpa1 (qtotal, CarryN2[10:3], SumN2[10:3]);
    // q = {+2, +1, -1, -2} else q = 0
    qst pd1 (quotient, qtotal[7:1], divi1[6:4]);
    or o1 (ulp, quotient[2], quotient[3]);
    mux41hx11 mux3 (mdivi_temp, divi2c, divi1c, divi1,
                   divi2, quotient);
    nor n1 (zero, quotient[3], quotient[2], quotient[1],
            quotient[0]);
    mux21x11 mux4 (mdivi, mdivi_temp, 11'b000000000, zero);
    csa11 csa1 (Sum, Carry, mdivi, SumN2,
                {CarryN2[10:1], ulp});

endmodule // divide4
```

Figure 5.10. Radix 4 Verilog Code.

To reduce the delay, the second radix 4 utilizes a quotient digit set that is computed conditionally (i.e. they are formed from the truncated shifted partial remainder). The quotient digit is determined each iteration first by selecting q_H and then using this value to choose q_L based on truncated shifted partial remainders. In order to get the shifted partial remainders for q_L, the lower 9 bits are utilized of the 11 bit partial remainder as shown in Figure 5.11 (since this would be the shifted partial remainder on the following iteration). The block diagram of the new QST is shown in Figure 5.12. Notice that both q_{i+1} and q_{i+2} are generated in one block.

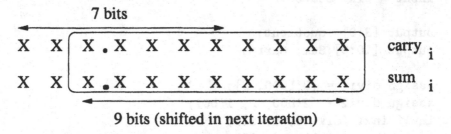

Figure 5.11.   Choosing Correct Residuals.

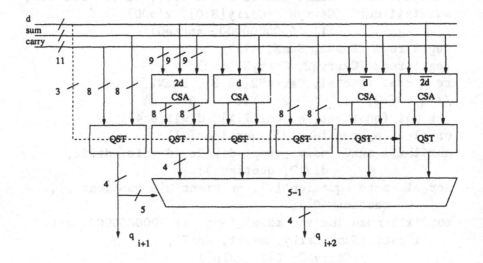

Figure 5.12.   Quotient Digit Selection for Radix 16 Division.

For an easy implementation, the same QST function is utilized as the radix 4 implementation. However, careful attention must be given so that the conditional computation should be correct, therefore, one extra bit is examined. Therefore, a 9 bit CSA is utilized to add the partial remainders in carry-save format to the appropriate shifted divisor. The quotient bit is determined during each iteration by selecting the quotient from a conditional residuals. Similar to the radix 4 implementation a CPA is utilized to reduce the size of the QST. Consequently, 8 most-significant bits of the carry-save output are input into a CPA as shown in Figure 5.12. Once q_L is known, it is utilized to select the

```
module qst2 (zero1, zero2, qj, qjn,
    divi2, divi1, divi1c, divi2c, carry, sum);

    input [10:0] carry, sum;
    input [10:0] divi2, divi1, divi1c, divi2c;

    output [3:0] qj;
    output [3:0] qjn;
    output  zero1, zero2;

    csa9 add1 (d2s, d2c, divi2[8:0], sum[8:0], carry[8:0]);
    csa9 add2 (d1s, d1c, divi1[8:0], sum[8:0], carry[8:0]);
    csa9 add3 (d1cs, d1cc, divi1c[8:0], sum[8:0], carry[8:0]);
    csa9 add4 (d2cs, d2cc, divi2c[8:0], sum[8:0], carry[8:0]);
    rca8 cpa1 (qtotal, sum[10:3], carry[10:3]);
    rca8 cpa2 (qt2, d2s[8:1], d2c[8:1]);
    rca8 cpa3 (qt1, d1s[8:1], d1c[8:1]);
    rca8 cpa4 (qt0, sum[8:1], carry[8:1]);
    rca8 cpa5 (qt1c, d1cs[8:1], d1cc[8:1]);
    rca8 cpa6 (qt2c, d2cs[8:1], d2cc[8:1]);
    qst pd1 (qj, qtotal[7:1], divi1[6:4]);
    nor n1 (zero1, qj[3], qj[2], qj[1], qj[0]);
    qst pd2 (q2, qt2[7:1], divi1[6:4]);
    qst pd3 (q1, qt1[7:1], divi1[6:4]);
    qst pd4 (q0, qt0[7:1], divi1[6:4]);
    qst pd5 (q1c, qt1c[7:1], divi1[6:4]);
    qst pd6 (q2c, qt2c[7:1], divi1[6:4]);
    nor n2 (zero2, qjn[3], qjn[2], qjn[1], qjn[0]);
    mux51hx4 mux1 (qjn, q2c, q1c, q1, q2, q0, {qj, zero1});

endmodule // qst2
```

Figure 5.13. Quotient Digit Selection for Radix 16 Division Verilog Code.

correct version of q_H from all the possible combinations of q_H. The Verilog code for the radix 16 QST is shown in Figure 5.13. Both *zero*1 and *zero*2 are utilized to decode a 0 for q_{i+1} and q_{i+2}, respectively.

The implementation of radix 16 using overlapping stages is quite efficient without increasing the hardware too dramatically. The block diagram of the radix 16 datapath is shown in Figure 5.14. For this implementation, only one set of registers to store the carry and sum parts for the second radix 4 stage are utilized. Therefore, each iteration retires 4 bits of the quotient improving the performance. The delay would theoretically correspond to two times the delay of a radix 4 implementation. The worst-case delay is illustrated by the dotted line. The Verilog code is shown in Figure 5.15. As in previous implementations, both on-the-fly conversion and zero detection are not shown.

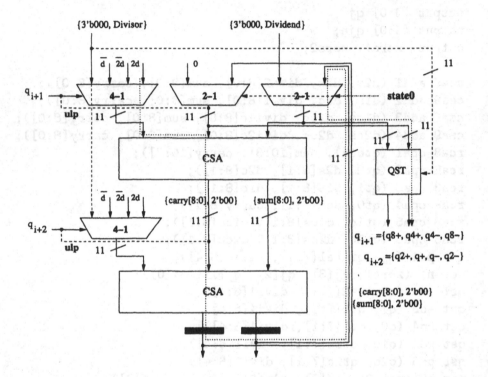

Figure 5.14. Radix 16 Division.

5.8   Summary

Division by recurrence can be quite efficient provide the QST is chosen wisely. There are several other variations on recurrence division including scaling the operands, prediction, and very high radix division [EL94]. The basic recurrence for square root is conceptually similar to divide as:

$$w_{i+1} = r \cdot w_i - 2 \cdot S_i \cdot s_{i+1} - s_{i+1}^2 \cdot r^{(-i+1)}$$

As can be seen by this recurrence, it is somewhat more complex than division. However, because of this similarity, square root can be implemented as a combined divide and square root unit.

```
module divide16 (qj, qjn, Sum1, Carry1, Sum2, Carry2,
        op1, op2, clk, state0);

   input [7:0]  op1, op2;
   input   clk, state0;

   output [3:0]  qj, qjn;
   output [10:0] Sum1, Carry1;
   output [10:0] Sum2, Carry2;

   assign divi1 = {3'b000, op2};
   assign divi2 = {2'b00, op2, 1'b0};
   inv11 inv1 (divi1c, divi1);
   inv11 inv2 (divi2c, divi2);
   assign dive1 = {3'b000, op1};
   mux21x11 mux1 (SumN, {Sum2[8:0], 2'b00}, dive1, state0);
   mux21x11 mux2 (CarryN, {Carry2[8:0], 2'b00},
                  11'b00000000000, state0);
   reg11 reg1 (SumN2, SumN, clk);
   reg11 reg2 (CarryN2, CarryN, clk);
   // quotient = {+2, +1, -1, -2} else q = 0
   qst2 pd1 (zero1, zero2, qj, qjn,
     divi2, divi1, divi1c, divi2c, CarryN2, SumN2);
   or o1 (ulp1, qj[2], qj[3]);
   or o2 (ulp2, qjn[2], qjn[3]);
   mux41hx11 mux3 (mdivi1_temp, divi2c, divi1c, divi1,
                  divi2, qj);
   mux41hx11 mux4 (mdivi2_temp, divi2c, divi1c, divi1,
                  divi2, qjn);
   mux21x11 mux5 (mdivi1. mdivi1_temp, 11'b00000000000,
                  zero1);
   mux21x11 mux6 (mdivi2, mdivi2_temp, 11'b00000000000,
                  zero2);
   csa11 csa1 (Sum1, Carry1, mdivi1, SumN2,
               {CarryN2[10:1], ulp1});
   csa11 csa2 (Sum2, Carry2, mdivi2,
       {Sum1[8:0], 2'b00}, {Carry1[8:0], 1'b0, ulp2});

endmodule // divide16
```

Figure 5.15.   Radix 16 Division Verilog Code.

Chapter 6

ELEMENTARY FUNCTIONS

Elementary functions are one of the most challenging and interesting arithmetic computations available today. Many of the advancements in this area result from mathematical theory that has spanned many centuries of work. Elementary functions are typically the referred to as the most commonly utilized mathematical functions such as *sin*, *cos*, *tan*, exponentials, and logarithms [Mul97].

Since elementary functions are implemented in digital arithmetic, they are prone to error that is introduced due to rounding and truncation [Gol91]. Therefore, it is important that when utilizing elementary functions that a proper error analysis be done before the actual implementation. This will guarantee that the result will guarantee the necessary accuracy required for the implementation. The reason this occurs is because elementary functions usually can not be computed exactly with a finite number of arithmetic operations.

As opposed to most implementations in this book, elementary functions usually utilize lookup tables to either compute the evaluation or assist in the computation. Consequently, many implementations can not be efficiently built in Silicon without memory generation. Therefore, even though the implementations shown here utilize memory for table lookups, many of the implementations can not be synthesized or assembled easily. The reason for this is that memory is composed of many different elements that are not straight forward to place such as column/row decoders and sense amplifiers. For smaller size tables, Silicon can be built with pull up and pull down elements, however, these implementations are not efficient as properly organized memory elements. On the other hand, custom-built devices can be built regardless of the implementation at the expense of design time.

Most implementations are usually broken into three categories. This is because elementary functions can theoretically implemented for a limited range. The steps involved in the processing are the following:

1 Preprocessing : Usually involves reducing the input operand. This transformation depends on the function and the method that is used for approximating the function.

2 Evaluation : Calculate the approximation.

3 Postprocessing : Usually involves reconstructing the approximation of the function

Of course, many of these steps listed above can be combined or eliminated if needed depending on the criteria for the implementation. Table 6.1 shows the input and output ranges for various elementary functions. Well-known techniques, such as those presented in [CW80] and [Wal71], can be used to bring the input operands within the specified range.

$f(x)$	Input Range	Output Range
$1/x$	$[1, 2)$	$(0.5, 1]$
$\sqrt{x}$	$[1, 2)$	$[1, \sqrt{2})$
$\sin(x)$	$[0, 1)$	$[0, 1/\sqrt{2})$
$\cos(x)$	$[0, 1)$	$(1/\sqrt{2}, 1]$
$\tan^{-1}(x)$	$[0, 1)$	$[0, \pi/4)$
$\log_2(x)$	$[1, 2)$	$[0, 1)$
2^x	$[0, 1)$	$[1, 2)$

Table 6.1. Ranges for Various Functions.

Elementary functions cannot be computed exactly within a finite number of operations [LMT98]. Elementary functions are usually evaluated as approximations to a specific function. This problem is sometimes called the Table Marker's dilemma which states that rounding these functions correctly at a reasonable cost is difficult. For a given rounding mode a function is said to be correctly rounded, if for any input operand, the result would be obtained if the evaluation of the function is first computed exactly with infinite precision and then rounded.

There are three classification of elementary functions. They are the following:

1 Table Lookup

2 Shift and Add

3 Polynomial Approximations

Only the first two type of elementary functions are presented in this chapter. A good reference for all types of elementary function generation can be found in [Mul97]. Since polynomial approximations can be implemented similar to table lookup or shift and add algorithms, they are not included to save space.

6.1 Generic Table Lookup

One of the fastest and simplest ways to implement elementary functions is by use of a table lookup. In other words, an operand is input into a simple memory element as an address with m bits and output with k bits. Therefore, the total size of the memory element is $2^m \times k$ bits. Inside this memory element is the evaluation of each input operand or $a(X_0)$. This is visualized in Figure 6.1 where m bits from an n bit register is utilized to initiate the lookup. A table lookup does not have to utilize the entire operand for a lookup. For example, in Figure 6.1 $n - m$ are not utilized. This is common in many implementations where the table lookup may be utilized in some sort of post-processing. When the initial table lookup is utilizing in subsequent hardware to make the approximation more accurate, it is called a table-driven method.

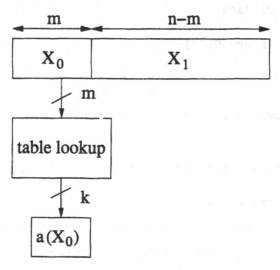

Figure 6.1. Table Lookup.

As stated earlier, many implementations of table lookup are done via memory compilers. Therefore, writing Verilog code is useful only for verification. There are some synthesis packages that are able to generate ROM or RAM elements, however, due to the variety of memory elements that are available, the task is best left to a memory compiler. On the other hand, table lookups can be implemented using ROMs, RAMs, PLA or combinational logic, but ROM tables are most common.

For example, suppose a 4-input operand is attempting to implement a table lookup implementation for sine. Assuming that the output is $m = 4$, each memory address will contain the sine of the input [IEE85]. Utilizing the ROM code found in Chapter 2, the Verilog is shown in Figure 6.2 where *table.dat* stores the values inside the table. The values for *table.dat* are shown in Figure 6.3. For example, if $x = 0.875 = 0.1110$, the evaluation of this is $sin(0.875) = 0.75 = 0.1100$ which is the 15th entry in the *table.dat* file since $0.875 = 0.1110 \rightarrow 1110 = 15$. Since $m = 4$, there are 16 possible values inside the data file called table.dat. Careful attention should be made to make sure that the data file has the correct number of data elements inside it. Most Verilog compilers may not complain about the size and either chop off or zero out memory elements if the size is not correct.

```
module tablelookup (data,address);

    input  [3:0] address;
    output [3:0] data;

    reg [3:0] memory[0:31];

    initial
      begin
        $readmemb("./table.dat", memory);
      end

    assign data = memory[address];

endmodule // tablelookup
```

Figure 6.2. Table Lookup Verilog Code.

```
0000
0001
0010
0011
0100
0101
0110
0111
1000
1001
1001
1010
1011
1100
1100
1101
```

Figure 6.3. Table Lookup Verilog Code.

6.2 Constant Approximations

Sometimes designers do not have the resources to implement memory elements for elementary functions. This is very common since memory elements tend to consume a large amount of memory. One method that is useful both in software and hardware are constant approximations. For constant approximations, a constant provides an approximation to a function on an interval $[a, b]$. For example, a reciprocal function (i.e. $1/x$) is shown in Figure 6.4. The constant, a_0 approximates the function over $[a, b]$.

To minimize the maximum absolute error, a_0 can be approximated as

$$a_0 = \frac{\frac{1}{a} + \frac{1}{b}}{2} = \frac{a + b}{2 \cdot a \cdot b}$$

Since utilizing a constant for the approximation introduces an error, it is important to understand what this error is. The maximum absolute error is

$$\epsilon = |\frac{1}{a} - a_0| = |\frac{1}{a} - \frac{a + b}{2 \cdot a \cdot b}| = |\frac{b - a}{2 \cdot a \cdot b}|$$

For example, to approximate $f(X) = 1/X$ for X on the interval $[1,2]$, $a_0 = (1/2 + 1)/2 = 0.75$ and $\epsilon = |(1 - 1/2)/2| = 0.25$, which means the approximation is accurate to 2 bits. In general, a_0 may need to be rounded to k fractional bits, which results in a maximum absolute error of 2^{-k-1}.

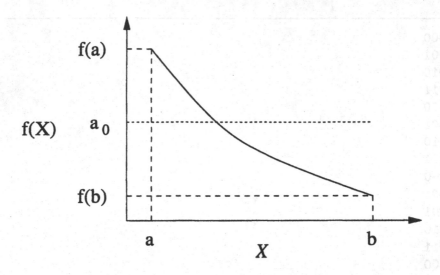

Figure 6.4. Constant Approximation.

To implement this in Verilog is quite simple. The best way is to utilize the *assign* keyword and the actual bit value. For example, to implement the value 0.75 up to 5 bits would be implemented as follows:

```
assign A = 5'b01100;
```

The radix point is imaginary and assumed to be at bit position 4. Usually, synthesis packages will have not have any problem implementing this code because it usually can associate this value with either a pull-up or pull-down logic [KA97]. If pull-up or pull-down logic is not available, the logic is connected to VDD or GND, respectively.

6.3   Piecewise Constant Approximation

To improve the accuracy of the constant approximation, the interval $[a, b]$ can be sub-divided into sub-intervals. This is called a piecewise constant approximation. Piecewise Constant Approximations break the input operand into two parts: a m-bit most-significant part, X_0, and a $(n - m)$-bit least-significant part, X_1, as in Figure 6.1. The bits from X_0 are used to perform a table lookup, which provides an initial approximation $a_0(X_0)$. If m fractional bits are utilized for the table lookup, and the output from the table is k bits, there are 2^m sub-intervals where the size of each sub-interval is 2^{-m} (i.e. an ulp). On the sub-interval, $[a + i \cdot 2^{-m}, a + (i + 1) \cdot 2^{-m}]$, the coefficient is selected as

$$a_0(X_0) = \frac{f(a + i \cdot 2^{-m}) + f(a + (i + 1) \cdot 2^{-m})}{2}$$

using the formula from the previous section. The maximum error is

$$\epsilon = \left| \frac{f(a + i \cdot 2^{-m}) - f(a + (i+1) \cdot 2^{-m})}{2} \right|$$

Figure 6.5 shows a visual representation of piecewise constant approximation utilizing 2 sub-intervals. For example, on the interval $[0.5, 0.625]$, the constant would be

$$a_0(X_0) = \frac{0.5 + 0.625}{2 \cdot 0.5 \cdot 0.625} = 1.8$$

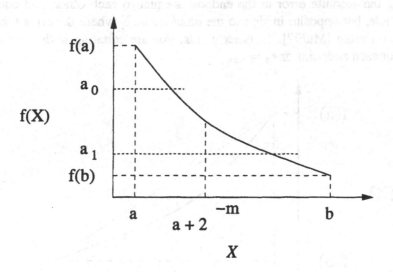

Figure 6.5. Piecewise Constant Approximation.

The implementation for this method is similar to the constant approximation. The implementation can also be implemented using the *assign* keyword and a multiplexor. For example, the following statement implements a piecewise constant approximation based on one bit of d:

```
mux21 mux_ia(ia_out, 8'b00110000, 8'b11010000, d);
```

Certain functions like the ones in Table 6.1 do not to be stored. Therefore, these values can be removed from the table. For example, when approximating $1/X$ for X on $[1, 2]$, the leading ones in X and $1/X$ do not need to be stored in the table.

6.4   Linear Approximations

Another popular form of approximation is the linear approximations. Linear approximations are better than constant approximations, however, there is more complexity in the implementation. Linear approximations provide an approximation based on a first order approximation:

$$f(X) \approx c_1 \cdot X + c_0$$

The error is computed by the following equation:

$$\epsilon = f(X) - (c_1 \cdot X + c_1)$$

This is shown in Figure 6.6. The maximum absolute error is minimized by making the absolute error at the endpoints equal to each other, and equal in magnitude, but opposite in sign to the point on $[a, b]$ where the error takes its maximum value [Mul97]. In other words, you are equating both error equations for each endpoint or $\epsilon_a = \epsilon_b$.

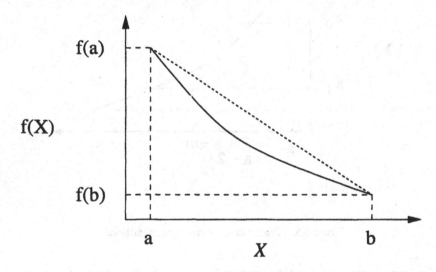

Figure 6.6.   Linear Approximation.

For example, assume that the function we wish to approximate is the reciprocal, $f(X) = 1/X$. Then,

$$\frac{1}{a} - c_1 \cdot a - c_0 = \frac{1}{b} - c_1 \cdot b - c_0$$

$$\frac{1}{a} - \frac{1}{b} = c_1 \cdot (a - b)$$

$$\frac{b - a}{a \cdot b} = c_1 \cdot (a - b)$$

$$c_1 = \frac{-1}{a \cdot b}$$

To find the y-intercept its important to set the input to the maximum value (i.e. since this minimizes the maximum error) which can be computed by taking the derivative with respect to the input operand and setting it to 0. For example, with the reciprocal

$$\epsilon = \frac{1}{X_{max}} - (c_1 \cdot X_{max} + c_0)$$

$$\frac{\partial \epsilon}{\partial X_{max}} = \frac{-1}{X_{max}^2} - c_1 = 0$$

$$X_{max} = \frac{-1}{\sqrt{-c_1}}$$

$$X_{max} = \sqrt{a \cdot b}$$

Letting $\epsilon(a) = \epsilon(b) = -\epsilon(\sqrt{a \cdot b})$ will enable solving for the y-intercept. From this equation, either a or b can be utilized to solve the equation. Therefore,

$$\frac{1}{a} - \frac{-1}{a \cdot b} \cdot a - c_0 = -\left(\frac{1}{\sqrt{a \cdot b}} - \frac{-1}{a \cdot b} \cdot \sqrt{a \cdot b} - c_0 \right)$$

$$\frac{1}{a} + \frac{1}{b} - c_0 = -\left(\frac{2}{\sqrt{a \cdot b}} - c_0 \right)$$

$$2 \cdot c_0 = \frac{a + b}{a \cdot b} + \frac{2}{\sqrt{a \cdot b}}$$

$$c_0 = \frac{a + b}{2 \cdot a \cdot b} + \frac{1}{\sqrt{a \cdot b}}$$

$$c_0 = \frac{a + \sqrt{a \cdot b} + b}{2 \cdot a \cdot b}$$

For example, for reciprocal and $X = [1, 2]$, $c_0 = 1.45710678118655$ and $c_1 = -0.5$. With the use of the Maple numerical software package its easy to check this with the *minimax* function [CGaLL+91].

```
with(numapprox);
minimax(1/x, x=1..2, 1);
1.457106781 - .5000000000 x
```

A plot of the linear approximation versus actual is shown in Figure 6.7.

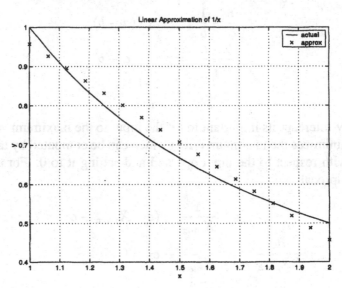

Figure 6.7. Plot of Linear Approximation.

6.4.1   Round to Nearest Even

Although error analysis is a valuable method of analyzing an algorithm's implementation, careful attention should be paid to the size of each bit pattern. Most error analysis comes to the conclusion that for achieving a certain size or accuracy requires precision that is greater or equal to the required precision. This necessitates some sort of method to round the answer to its desired precision. For example, typical floating-point dividers typically compute approximately 58 bits of precision when they only need 53 bits for the final quotient.

One method to convert a result which is greater than the required precision is by rounding. The most popular method of rounding is called round-to-nearest-even (RNE) that is part of the IEEE-754 floating-point standard [IEE85], [Sch03]. To perform round-to-nearest even (RNE), it is important to know where the radix point is located so that an ulp can be added and what the error analysis is.

In Table 6.2 only two bits (i.e. the G and R bit) are utilized to perform RNE, however, this can be augmented or decremented accordingly. Table 6.2 docu-

Number (XLGR)	Rounded Value	Error Rounded Value - Number	Round
X0.00	X0.	0	0
X0.01	X0.	-1/4	0
X0.10	X0.	-1/2	0
X0.11	X0. + ulp	+1/4	1
X1.00	X1.	0	0
X1.01	X1.	-1/4	0
X1.10	X1. + ulp	+1/2	1
X1.11	X1. + ulp	+1/4	1
Total Error		0.0	

Table 6.2. Round-to-Nearest-Even Scheme

ments when an ulp is to be added to the least significant bit (i.e. an ulp) where L is the least significant bit before the rounding point, G is the guard digit used for preventing catastrophic cancellation [Hig94], and R is the rounding bit [Mat87]. For RNE, the main advantage is that it makes the total average error 0.00. The main difference between general rounding and RNE is the case when the value to be rounded is exactly 0.500000.... For the case where the value to be rounded is exactly 0.5 (i.e. $0.5 = 0.10$ from Table 6.2), rounding occurs only if the L bit is 1. The *Round* column in Table 6.2 indicates whether rounding should occur. For this reason, the zero detect circuit is utilized with division implementations in Chapter 5. Therefore, the logic for RNE by examining the L, G, and R bit is

```
assign ulp = (a[6]&a[5] | a[5]&a[4]);
```

The Verilog code for a simple RNE implementation is shown in Figure 6.8. For this code, bit 8 is L, bit 7 is G, and bit 6 is R. Several recent designs have been able to modify prefix addition to support rounding efficiently [Bur02].

Therefore, the implementation will involve a multiplier and storage of the slope and y-intercept. The storage can easily done via an *assign* statement and utilizing a csam from Chapter 4, Figure 6.9 shows the Verilog implementation for 8 bits. Round-to-nearest even rounding is utilized to round the product within the multiplier to 8 bits. The instantiated multiplier, *csam8*, is a two's complement multiplier since the constants utilize positive and negative values. It is also extremely important to examine where the radix point is in fractional arithmetic as discussed in Chapter 4. Therefore, careful attention is made to make sure that the rounding point is observed within the *rne* instantiation.

```
module rne (z, a);

    input [15:0]  a;

    output [7:0]  z;

    assign ulp = (a[8]&a[7] | a[7]&a[6]);
    ha ha1 (z[0], w0, a[8], ulp);
    ha ha2 (z[1], w1, a[9], w0);
    ha ha3 (z[2], w2, a[10], w1);
    ha ha4 (z[3], w3, a[11], w2);
    ha ha5 (z[4], w4, a[12], w3);
    ha ha6 (z[5], w5, a[13], w4);
    ha ha7 (z[6], w6, a[14], w5);
    ha ha8 (z[7], w7, a[15], w6);

endmodule // mha8
```

Figure 6.8.   RNE Verilog Code.

```
module linearapprox (Z, X);

    input  [7:0] X;
    output [7:0] Z;

    assign C0 = 8'b01011101;
    assign C1 = 8'b11100000;

    csam8 mult1 (Zm, C1, X);
    rne round1 (Zt, Zm);
    rca8 cpa1 (Z, Zt, C0);

endmodule // linearapprox
```

Figure 6.9.   Linear Approximation Verilog Code.

The maximum absolute error is then computed as

$$
\begin{aligned}
\epsilon &= \frac{1}{a} - c_1 \cdot a - c_0 \\
&= \frac{1}{a} - \frac{-1}{a \cdot b} \cdot a - \left(\frac{a+b}{2 \cdot a \cdot b} + \frac{1}{\sqrt{a \cdot b}} \right) \\
&= \frac{1}{a} + \frac{b}{-2 \cdot a \cdot b} \frac{a+b}{\sqrt{a \cdot b}} - \frac{1}{\sqrt{a \cdot b}} \\
&= \frac{a+b}{a \cdot b} - \frac{a+b}{2 \cdot a \cdot b} - \frac{1}{\sqrt{a \cdot b}} \\
&= \frac{2 \cdot b + 2 \cdot a - (a+b) - 2 \cdot \sqrt{a \cdot b}}{2 \cdot a \cdot b} \\
&= \frac{a+b - 2 \cdot \sqrt{a \cdot b}}{2 \cdot a \cdot b}
\end{aligned}
$$

For the reciprocal function and an input range $[1, 2]$, the maximum absolute error is a little more than 4 bits. This is better than a constant approximation, however, it involves more complexity for the hardware. To improve the error in the approximation a table lookup can be used to provide a piecewise linear approximation.

6.5 Bipartite Table Methods

For applications that require low-precision elementary function approximations at high speeds, table lookups are often employed. However, as the required precision of the approximation increases, the size of the memory needed to implement the table lookups becomes prohibitive. Recent research into the design of bipartite tables has significantly reduced the amount of memory needed for high-speed elementary function approximations [HT95], [SM95]. With bipartite tables, two table lookups are performed in parallel, and then their outputs are added together [SS99], [SS98]. Extra delay is incurred due to the addition, however, it is minimal compared to the advantage gained by the reduction in memory.

To approximate a function using bipartite tables, the input operand is separated into three parts. The three partitions are denoted as x_0, x_1, and x_2 and have lengths of n_0, n_1, and n_2, respectively. The value of the input operand is $x = x_0 + x_1 + x_2$ and it has a length of $n = n_0 + n_1 + n_2$. The function is approximated as

$$
f(x) \approx a_0(x_0, x_1) + a_1(x_0, x_2)
$$

The $n_0 + n_1$ most significant bits of x are inputs to a table that provides the coefficient $a_0(x_0, x_1)$, and the n_0 most significant and n_2 least significant bits

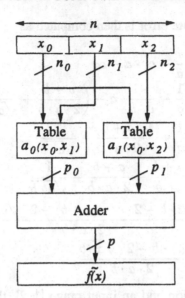

Figure 6.10.   Bipartite Table Method Block diagram (Adapted from [SS98]).

of x are inputs to a table that provides the coefficient $a_1(x_0, x_2)$. The outputs from the two tables are summed to produce an approximation $\widetilde{f}(x)$. as shown in Figure 6.10

This technique is extended by partitioning x into $m+1$ parts, $x_0, x_1, \ldots, x_m$, with lengths of $n_0, n_1, \ldots, n_m$, respectively. This approximation takes the form

$$f(x) \approx \sum_{i=1}^{m} a_{i-1}(x_0, x_i)$$

The hardware implementation of this method has m parallel table lookups followed by an m-input multi-operand adder. The ith table takes as inputs x_0 and x_i. The sum of the outputs from the tables produces an approximation to $f(x)$. By dividing the input operand into a larger number of partitions, smaller tables are needed.

6.5.1   SBTM and STAM

The approximations for the Symmetric Bipartite Table Method (SBTM) and Symmetric Table Addition Method (STAM) are based on Taylor series expansions centered about the point $x_0 + x_1 + \delta_2$. The SBTM uses symmetry in the entries of one of the tables to reduce the overall memory requirements. The value

$$\delta_2 = 2^{-n_0-n_1-1} - 2^{-n_0-n_1-n_2-1}$$

is exactly halfway between the minimum and maximum values for x_2. Using the first two terms of the Taylor series results in the following approximation

$$
\begin{aligned}
f(x) &\approx f(x_0 + x_1 + \delta_2) \\
&+ f'(x_0 + x_1 + \delta_2)(x_2 - \delta_2)
\end{aligned}
$$

As discussed in [SS99], the omission of the higher order terms in the Taylor series leads to a small approximation error. The first coefficient is selected as the first term in the Taylor series expansion.

$$
a_0(x_0, x_1) = f(x_0 + x_1 + \delta_2)
$$

Having the second coefficient depend on x_0, x_1, and x_2 would make the SBTM impractical due to its memory size. Thus, the second coefficient is selected as

$$
a_1(x_0, x_2) = f'(x_0 + \delta_1 + \delta_2)(x_2 - \delta_2)
$$

This corresponds to the second term of the Taylor series with x_1 replaced by the constant δ_1, where

$$
\delta_1 = 2^{-n_0-1} - 2^{-n_0-n_1-1}
$$

is exactly halfway between the minimum and maximum values for x_1.

One benefit of this method is that the magnitude of the second coefficient is substantially less than the first coefficient, which allows the width of the second table to be reduced. Since $\mid x_2 - \delta_2 \mid < 2^{-n_0-n_1-1}$, the magnitude of the second coefficient is bounded by

$$
\mid a_1(x_0, x_2) \mid < \mid f'(\xi_1) \mid 2^{-n_0-n_1-1}
$$

where ξ_i is the point at which $\mid f^i(x) \mid$ takes its maximum value. This results in approximately

$$
n_0 + n_1 + 1 + \log_2(\mid f(\xi_0)/f'(\xi_1) \mid)
$$

leading zeros (or leading ones if $a_1(x_0, x_2) < 0$). These leading zeros (or ones) are not stored in memory, but are obtained by sign-extending the most significant bit of $a_1(x_0, x_2)$ before performing the carry propagate addition.

Similar to the SBTM, the coefficients for the STAM are generated so that they have a large number of leading zeros. Although the STAM requires more tables than the SBTM, the size of each table and the total memory size is reduced. The number of inputs to the adder, however, is increased. It is assumed that $0 \leq x < 1$, and thus

$$
\begin{aligned}
0 \leq x_0 &\leq 1 - 2^{-n_0} \\
0 \leq x_i &\leq 2^{-n_{0:i}-1} - 2^{-n_{0:i}} \quad (1 \leq i \leq m)
\end{aligned}
$$

where $n_{j:k} = \sum_{i=j}^{k} n_i$.

To reduce the approximation error and create symmetry in the table coefficients, δ_i is defined to be exactly halfway between the minimum and maximum values of x_i, which gives

$$\delta_i = 2^{-n_{0:i-1}-1} - 2^{-n_{0:i}-1} \quad (1 \leq i \leq m)$$

It should be noted that δ_2 from the SBTM is equivalent to $\delta_{2:m}$ from the STAM, and x_2 from the SBTM is equivalent to $x_{2:m}$ from the STAM (i.e. $m = 2$).

The two term Taylor series expansion of $f(x)$ about $x_0 + x_1 + \delta_{2:m}$ is

$$\begin{aligned} f(x) &\approx f(x_0 + x_1 + \delta_{2:m}) \\ &+ f'(x_0 + x_1 + \delta_{2:m})(x_{2:m} - \delta_{2:m}) \end{aligned}$$

Similar to SBTM, x_1 is replaced by δ_1 which gives

$$\begin{aligned} f(x) &\approx f(x_0 + x_1 + \delta_{2:m}) \\ &+ f'(x_0 + \delta_1 + \delta_{2:m})(x_{2:m} - \delta_{2:m}) \end{aligned}$$

The second term in this approximation is then distributed into $m - 1$ terms, which gives

$$\begin{aligned} f(x) &\approx f(x_0 + x_1 + \delta_{2:m}) \\ &+ f'(x_0 + \delta_1 + \delta_{2:m}) \cdot \sum_{i=2}^{m}(x_i - \delta_i) \end{aligned}$$

Thus, the values for the coefficients are

$$\begin{aligned} a_0(x_0, x_1) &= f(x_0 + x_1 + \delta_{2:m}) \\ a_{i-1}(x_0, x_i) &= f'(x_0 + \delta_1 + \delta_{2:m})(x_i - \delta_i) \end{aligned}$$

where $2 \leq i \leq m$. It is equivalent for SBTM if $m = 2$ and STAM if $m > 2$. The number of leading zeros (or ones) in $a_{i-1}(x_0, x_i)$ is determined by the bound

$$\mid a_{i-1}(x_0, x_i) \mid < \mid f'(\xi_1) \mid 2^{-n_{0:i}-1}$$

The table for $a_i(x_0, x_i)$ has $2^{n_0 + n_i}$ words. The method for selecting the coefficients causes tables $a_1(x_0, x_2)$ through $a_{m-1}(x_0, x_m)$ to be symmetric and allows their sizes to be reduced to $2^{n_0 + n_i - 1}$ words. The symmetry achieved in the SBTM and the STAM is obtained because

- $2\delta_i - x_i$ is the one's complement of x_i

- $a_{i-1}(x_0, 2\delta_i - x_i)$ is the one's complement of $a_{i-1}(x_0, x_i)$

These properties are demonstrated in Table 6.3 which gives eight table entries for $f(x) = cos(x)$ when $m = 2$, $n_0 = 2$, $n_1 = 2$, $n_2 = 3$, $g = 2$, and $p_f = 7$. Each value for x_2 in the first half of the table has a one's complement in the second half of the table, as shown by the bold-faced binary values. The corresponding table entries for $a_1(x_0, x_2)$ also have one's complements, shown in bold. There is no need to store the bits of $a_1(x_0, x_2)$ which are not bold-faced, since these correspond to leading zeros (or ones) that can be obtained by sign-extension.

x		$a_1(x_0, x_2)$	
decimal	binary	decimal	binary
0.500000	0.100**0000**	+0.0166016	0.00000**10001**
0.507812	0.100**0001**	+0.0107422	0.000000**1011**
0.515625	0.100**0010**	+0.0068359	0.0000000**111**
0.523438	0.100**0011**	+0.0029297	0.00000000**11**
0.531250	0.100**0100**	-0.0029297	1.11111111**01**
0.539062	0.100**0101**	-0.0068359	1.1111111**001**
0.546875	0.100**0110**	-0.0107422	1.111111**0101**
0.554688	0.100**0111**	-0.0166016	1.11111**01111**

Table 6.3. Table Entries for $a_1(x_0, x_2)$ (Adapted from [SS98]).

With the SBTM the $a_1(x_0, x_2)$ table is folded by examining the most significant bit of x_2. If this bit is a zero, then the remaining bits of x_2 remain unchanged, and the value is read from the $a_1(x_0, x_2)$ table and added to the $a_0(x_0, x_1)$ table. However, if this bit is a one, then the remaining bits of x_2 are complemented, and used to address the a_1 table. The output is then complemented and added to $a_0(x_0, x_1)$. This method is extended for the STAM by examining the most significant bit of x_i ($2 \leq i \leq m$), as shown in Figure 6.11

If the most significant bit of x_i is a one, a row of $n_i - 1$ XOR gates complements the remaining bits of x_i which addresses the $a_{i-1}(x_0, x_i)$ table, and the output of the table is complemented using a row of p_i exclusive-or gates. The most significant bits and the least significant bit of the table do not need to be stored, since these bits are known in advance.

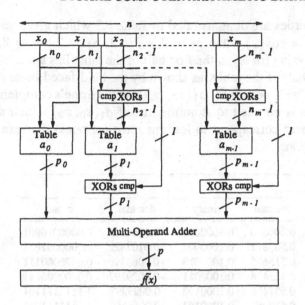

Figure 6.11.   Generalized Table Addition Method Block Diagram (Adapted from [SS98]).

The coefficients and the final result are rounded using a method similar to the one described in [SM93]. If the final result has p_f fraction bits, and the coefficients each have $(p_f + g + 1)$ fraction bits, rounding is performed as follows:

- If m is even, $a_0(x_0, x_1)$ is rounded to the nearest number with $(p_f + g)$ fraction bits and the least significant bit is set to zero. Otherwise, $a_0(x_0, x_1)$ is truncated to $(p_f + g)$ fraction bits and the least significant bit is set to one.

- For $i > 1$, $a_{i-1}(x_0, x_i)$ is truncated to $(p_f + g)$ fraction bits and the least significant bit is set to one.

- $\widetilde{f(x)}$ is rounded to the nearest number with p_f fraction bits.

This method guarantees that the maximum absolute error in rounding each coefficient is bounded by 2^{-p_f-g-1}, and that the least significant bit of their sum is always a one.

By careful choosing the partitioning for SBTM and STAM, the implementation controls the errors and produces results that are faithfully rounded (i.e. the computed result differs from the true result by less than one unit in the last place (ulp) [SM93]). Faithful rounding is guaranteed if the following two conditions are met

$$2n_0 + n_1 \geq p_f + \log_2(|\, f''(\xi_2)\,|)$$
$$g \geq 2 + \log_2(m - 1)$$

For example, assume the goal is to approximate the sine function with the SBTM to $p_f = 12$ fraction bits on $[0, 1)$. If x is partitioned with $n_0 = 4$, $n_1 = 4$, and $n_2 = 4$, then $8 + 4 \geq 12 + \lceil \log_2(1/\sqrt{2}) \rceil$ satisfies the first inequality.

Figure 6.12 shows a implementation of SBTM for reciprocal. The input operand is 8 bits, however, the leading bit is always a 1, therefore, only 7 bits are input into the table. Moreover, the leading one inside the $a_0(x_0, x_1)$ table is also not stored. The partitioning utilized is $n_0 = 3$, $n_1 = 2$, $n_2 = 2$, and $g = 2$ which satisfies the error needed for a 7 bits output. The $a_0(x_0, x_1)$ data file (32 x 8) is shown in Figure 6.13, whereas, the $a_1(x_0, x_2)$ data file (3 x 16) is shown in Figure 6.14. The total size is $32 \cdot 8 + 3 \cdot 16 = 304$ bits as opposed to $2^7 \times 7 = 896$ bits for a conventional table lookup (compression = 2.95). The compression is the amount of memory required by a standard table lookup divided by the amount of memory required by the method being examined [SM93].

```
module sbtm (ia_out, ia_in);

    input [6:0]    ia_in;

    output [7:0] ia_out;

    romia0 r0(rom0out, ia_in[6:2]);
    assign p0 = {1'b1, rom0out, 1'b0};
    xor x0(x0out, ia_in[1], ia_in[0]);
    romia1 r1(rom1out, {ia_in[6:4], x0out});
    xor3 x3(x3out, rom1out, ia_in[1]);
    assign p1 = {{6{ia_in[1]}}, x3out, 1'b1};
    rca10 add10(sum, cout, p0, p1);
    assign ia_out = {1'b0, sum[9:3]};

endmodule // sbtm
```

Figure 6.12. SBTM Reciprocal Verilog Code.

6.6 Shift and Add: CORDIC

Another recursive formula is utilized for elementary functions. The resulting implementation is called COordinate Rotation DIgit Computer (CORDIC) [Vol59]. The CORDIC algorithm is based on the rotation of a vector in a plane. The ro-

11111010
11101011
11011101
11001111
11000010
10110110
10101011
10100000
10010110
10001100
10000011
01111010
01110001
01101001
01100001
01011010
01010011
01001100
01000101
00111111
00111001
00110011
00101101
00101000
00100011
00011110
00011001
00010100
00001111
00001011
00000111
00000011

Figure 6.13. SBTM Reciprocal for $a_0(x_0, x_1)$ Data File.

tation is based on examining Cartesian coordinates on a unit circle as:

$$x = M \cdot cos(\alpha + \beta) = a \cdot cos\beta - b \cdot sin\beta$$
$$y = M \cdot sin(\alpha + \beta) = a \cdot sin\beta - b \cdot cos\beta$$

101
001
100
001
011
001
010
000
010
000
010
000
001
000
001
000

Figure 6.14.   SBTM Reciprocal for $a_1(x_0, x_2)$ Data File.

where M is the modulus of the vector and α is the initial angle as shown by Figure 6.15. The CORDIC algorithm performs computations on a vector by performing small rotations in a recursive manner. Each rotation, called a pseudo rotation, is performed until the final angle is achieved or a result is zero. As shown by the equations listed above, the rotations require a multiplication. The CORDIC algorithm transform the equations above through trigonometric identities to only utilize addition, subtraction, and shifting.

The resulting iterations after a few transformations are

$$x_{i+1} = x_i - \sigma_i \cdot 2^{-i} \cdot y_i$$
$$y_{i+1} = y_i + \sigma_i \cdot 2^{-i} \cdot x_i$$
$$z_{i+1} = z_i - \sigma_i \cdot tan^{-1}(2^{-i})$$

CORDIC also employs a scaling factor:

$$K = \prod_{i=0}^{\infty}(1 + 2^{-2 \cdot j})^{1/2} \approx 1.6468$$

The CORDIC algorithm utilizes two different modes called *rotation* and *vectoring*. In *rotation* mode, an initial vector (a, b) is rotated by an angle β. This is one of the most popular modes since $x_0 = 1/K$ and $y_0 = 0$ produces the *sin* β and cos β. On the other hand, in the *vectoring* mode, an initial vector

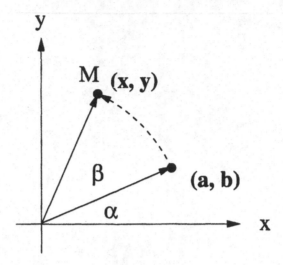

Figure 6.15.   Vector Rotation for CORDIC.

(a, b) is rotated until the b component is zero. The value of σ_{i+1} is chosen according the mode the CORDIC algorithm is currently in as shown in Table 6.4.

σ_{i+1}	Mode
$sign(Z_{i+1})$	Rotation
$sign(Y_{i+1})$	Vectoring

Table 6.4.   CORDIC Modes.

The block diagram of the CORDIC algorithm is shown in Figure 6.16. Two shifters are utilizes to shift the inputs based on the value of i. In addition, the table stores the values of $tan^{-1}(2^{-i})$ which also varies according to i. In order to implement CORDIC, three registers, indicated by the filled rectangles, are required to contain x_i, y_i, and z_i. The CPA is a controlled add or subtract similar to the RCAS in Chapter 3. Exclusive-or gates are inserted inside the CPA to select whether the blocks add or subtract. The values of σ_{i+1} are utilized to indicate to the CPA whether an addition or subtraction occurs based on the equations above.

The Verilog code for a 16-bit CORDIC code is shown in Figure 6.18. Figure 6.18 is written to compute the sine and cosine in rotation mode, however, it can easily be modified to handle both *rotation* and *vectoring* by adding a

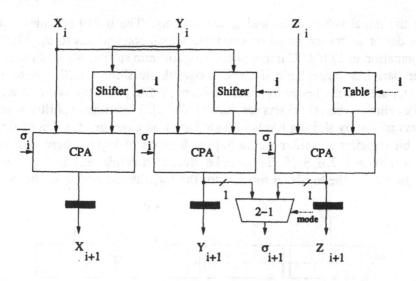

Figure 6.16. CORDIC Block Diagram.

2-1 multiplexor. The code has three main parts. One for the table that stores the value of $tan^{-1}(2^{-i})$. The table stores 16 values assuming 16 possible iterations could occur, however, varying the size of the table will determine the overall precision of the result [EL03], [Kor93]. The table values are based on rounded computations utilizing round-to-nearest even as shown here:

```
2D00
1A91
0E09
0720
0394
01CA
00E5
0073
0039
001D
000E
0007
0004
0002
0001
0000
```

The second part of the CORDIC code implements the two modules that shift and add is shown in Figure 6.19. The two constants, *constX* and *constY*, store the initial values of x_0 and y_0 respectively. The $inv16$ modules invoke the adder or subtractor based on inv input which corresponds to σ_i. This implementation of CORDIC implements a logarithmic shifter. Most shifters are either barrel or logarithmic shifters. Logarithmic shifters utilize powers of two and are usually better for larger shifters [WE85]. On the other hand, for smaller shifters, barrel shifters are better [WE85]. Logarithmic shifter work in powers of two by shifting levels of logic based on a power of 2. For example, a 32-bit logarithmic shifter would require 5 levels of logic, where each level would shift by $1, 2, 4, 8, 16$. This can be illustrated graphically in Figure 6.17. In Figure 6.21, the top-level module for the logarithmic shifter is shown and

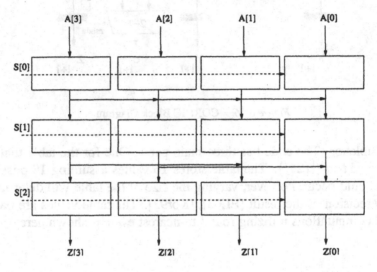

Figure 6.17. Logarithmic Shifter Block Diagram.

Figures 6.22, 6.23, 6.24, 6.25 detail the lower-level modules of the logarithmic shifter.

The third part of the CORDIC code implements the module that adds or subtracts the value of $tan^{-1}(2^{-i})$ as shown in Figure 6.20. All three modules have 16-bit registers to store the intermediate values every iteration. Although the implementation seems elaborate, it is quite efficient by using only addition, subtraction, or shifting as opposed to multiplication. Moreover, CORDIC can be modified for other trigonometric identities [Wal71].

6.7 Summary

This chapter presented several implementations for elementary functions. There are variety of algorithms available for implementations of elementary functions. The key factor to remember when implementing elementary functions is making sure the precision is accounted for by doing error analysis of

```
module cordic (sin, cos, data, currentangle,
       endangle, addr, load, clock);

   input [15:0]  endangle;
   input         clock;
   input [3:0]   addr;
   input         load;

   output [15:0] sin, cos;
   output [15:0] data, currentangle;

   angle angle1 (currentangle, load, endangle,
                 clock, data);
   sincos sincos1 (sin, cos, addr, load, clock,
                   currentangle[15]);
   rom lrom1 (data, addr);

endmodule // cordic
```

Figure 6.18. Main CORDIC Verilog Code.

the algorithm before the implementation starts. Polynomial approximations are not shown in this chapter, but can easily be implemented utilizing the techniques presented in this chapter. A polynomial approximation of degree n takes the form

$$f(X) \approx P_n(X) = \sum_{i=0}^{n} a_i \cdot X^i$$

```
module sincos (sin, cos, addr, load, clock, inv);

   input [3:0]   addr;
   input   inv, load, clock;

   output [15:0] sin;
   output [15:0] cos;

   assign constX=16'b0010011011011101;
   assign constY=16'b0000000000000000;

   mux21    m1 (invc, inv, 1'b0, load);

   mux21x16 m2 (outregX, cos, constX, load);
   shall log1 (outshX, outregX, addr);
   xor16 cmp1 (outshXb, outshX, invc);
   rca16 cpa1 (inregX, coutX, outregX, outshYb, invc);
   reg16 reg1 (cos, clock, inregX);

   mux21x16 m3 (outregY, sin, constY, load);
   shall log2 (outshY, outregY, addr);
   inv16 cmp2 (outshYb, outshY, ~invc);
   rca16 cpa2 (inregY, coutY, outregY, outshXb, ~invc);
   reg16 reg2 (sin, clock, inregY);

endmodule // sincos
```

Figure 6.19.   sincos CORDIC Verilog Code.

```
module angle (outreg, load, endangle, clock, data);

    input [15:0]  endangle;
    input   load, clock;
    input [15:0]  data;

    output [15:0] outreg;

    mux21x16 mux1 (inreg, outreg, endangle, load);
    xor16 cmp1 (subadd, data, ~inreg[15]);
    rca16 cpa1 (currentangle, cout, subadd, inreg,
                ~inreg[15]);
    reg16 reg1 (outreg, clock, currentangle);

endmodule // angle
```

Figure 6.20.   angle CORDIC Verilog Code.

```
module shall (dataout, a, sh);

    input [15:0]  a;
    input [3:0]   sh;

    output [15:0] dataout;

    logshift1 lev1 (l1, a, sh[0]);
    logshift2 lev2 (l2, l1, sh[1]);
    logshift4 lev3 (l4, l2, sh[2]);
    logshift8 lev4 (dataout, l4, sh[3]);

endmodule // shall
```

Figure 6.21.   Logarithmic Shifter Verilog Code.

```
module logshift1 (dataout, a, sh);

    input [15:0]  a;
    input    sh;

    output [15:0] dataout;

    mux21 m1 (dataout[0], a[0], a[1], sh);
    mux21 m2 (dataout[1], a[1], a[2], sh);
    mux21 m3 (dataout[2], a[2], a[3], sh);
    mux21 m4 (dataout[3], a[3], a[4], sh);
    mux21 m5 (dataout[4], a[4], a[5], sh);
    mux21 m6 (dataout[5], a[5], a[6], sh);
    mux21 m7 (dataout[6], a[6], a[7], sh);
    mux21 m8 (dataout[7], a[7], a[8], sh);
    mux21 m9 (dataout[8], a[8], a[9], sh);
    mux21 m10 (dataout[9], a[9], a[10], sh);
    mux21 m11 (dataout[10], a[10], a[11], sh);
    mux21 m12 (dataout[11], a[11], a[12], sh);
    mux21 m13 (dataout[12], a[12], a[13], sh);
    mux21 m14 (dataout[13], a[13], a[14], sh);
    mux21 m15 (dataout[14], a[14], a[15], sh);
    mux21 m16 (dataout[15], a[15], a[15], sh);

endmodule // logshift1
```

Figure 6.22.   logshift1 Shifter Verilog Code.

```
module logshift2 (dataout, a, sh);

   input [15:0]  a;
   input    sh;

   output [15:0] dataout;

   mux21 m1 (dataout[0], a[0], a[2], sh);
   mux21 m2 (dataout[1], a[1], a[3], sh);
   mux21 m3 (dataout[2], a[2], a[4], sh);
   mux21 m4 (dataout[3], a[3], a[5], sh);
   mux21 m5 (dataout[4], a[4], a[6], sh);
   mux21 m6 (dataout[5], a[5], a[7], sh);
   mux21 m7 (dataout[6], a[6], a[8], sh);
   mux21 m8 (dataout[7], a[7], a[9], sh);
   mux21 m9 (dataout[8], a[8], a[10], sh);
   mux21 m10 (dataout[9], a[9], a[11], sh);
   mux21 m11 (dataout[10], a[10], a[12], sh);
   mux21 m12 (dataout[11], a[11], a[13], sh);
   mux21 m13 (dataout[12], a[12], a[14], sh);
   mux21 m14 (dataout[13], a[13], a[15], sh);
   mux21 m15 (dataout[14], a[14], a[15], sh);
   mux21 m16 (dataout[15], a[15], a[15], sh);

endmodule
```

Figure 6.23. logshift2 Shifter Verilog Code.

```verilog
module logshift4 (dataout, a, sh);

    input [15:0]  a;
    input    sh;
    output [15:0] dataout;

    mux21 m1 (dataout[0], a[0], a[4], sh);
    mux21 m2 (dataout[1], a[1], a[5], sh);
    mux21 m3 (dataout[2], a[2], a[6], sh);
    mux21 m4 (dataout[3], a[3], a[7], sh);
    mux21 m5 (dataout[4], a[4], a[8], sh);
    mux21 m6 (dataout[5], a[5], a[9], sh);
    mux21 m7 (dataout[6], a[6], a[10], sh);
    mux21 m8 (dataout[7], a[7], a[11], sh);
    mux21 m9 (dataout[8], a[8], a[12], sh);
    mux21 m10 (dataout[9], a[9], a[13], sh);
    mux21 m11 (dataout[10], a[10], a[14], sh);
    mux21 m12 (dataout[11], a[11], a[15], sh);
    mux21 m13 (dataout[12], a[12], a[15], sh);
    mux21 m14 (dataout[13], a[13], a[15], sh);
    mux21 m15 (dataout[14], a[14], a[15], sh);
    mux21 m16 (dataout[15], a[15], a[15], sh);

endmodule // logshift4
```

Figure 6.24.   logshift4 Shifter Verilog Code.

```
module logshift8 (dataout, a, sh);

    input [15:0]  a;
    input    sh;

    output [15:0] dataout;

    mux21 m1 (dataout[0], a[0], a[8], sh);
    mux21 m2 (dataout[1], a[1], a[9], sh);
    mux21 m3 (dataout[2], a[2], a[10], sh);
    mux21 m4 (dataout[3], a[3], a[11], sh);
    mux21 m5 (dataout[4], a[4], a[12], sh);
    mux21 m6 (dataout[5], a[5], a[13], sh);
    mux21 m7 (dataout[6], a[6], a[14], sh);
    mux21 m8 (dataout[7], a[7], a[15], sh);
    mux21 m9 (dataout[8], a[8], a[15], sh);
    mux21 m10 (dataout[9], a[9], a[15], sh);
    mux21 m11 (dataout[10], a[10], a[15], sh);
    mux21 m12 (dataout[11], a[11], a[15], sh);
    mux21 m13 (dataout[12], a[12], a[15], sh);
    mux21 m14 (dataout[13], a[13], a[15], sh);
    mux21 m15 (dataout[14], a[14], a[15], sh);
    mux21 m16 (dataout[15], a[15], a[15], sh);

endmodule // logshift8
```

Figure 6.25. logshift8 Shifter Verilog Code.

Chapter 7

DIVISION USING MULTIPLICATIVE-BASED METHODS

This chapter presents methods for computing division by iteratively improving an initial approximation. Since this method utilizes a multiplication to compute divide it typically is called a multiplicative-divide method. Since multipliers occupy more area than addition or subtraction, the advantage to this method is that it provide quadratic convergence. Quadratic convergence means that number of bits of accuracy of the approximation doubles after each iteration. On the other hand, division utilizing recurrence methods only attains a linear convergence. In addition, many multiplicative-methods can be combined with multiplication function units making them attractive for many general-purpose architectures.

In this chapter, two implementations are shown with constant approximations. Both methods could be improved by inserting bipartite or other approximation methods to improve the initial estimate. Similar to division utilizing recurrence methods, multiplicative-methods can be modified to handle square root and inverse square root. Although these method obtain quadratic convergence, each method is only as good as its initial approximation. Therefore, obtaining an approximation that enables fast convergence to the operand size is crucial to making multiplicative-divide methods more advantageous than recurrence methods.

7.1 Newton-Raphson Method for Reciprocal Approximation

Newton-Raphson iteration is used to improve the approximation $X_i \approx 1/D$. Newton-Raphson iteration finds the zero of a function $f(X)$ by using the following iterative equation [Fly70].

$$X_{i+1} = X_i - \frac{f(X_i)}{f'(X_i)}$$

where X_i is an initial approximation to the root of the function, and X_{i+1} is an improvement to the initial approximation. To approximate $1/D$ using Newton-Raphson iteration, it is necessary to chose a function that is zero for $X = 1/D$ or

$$\begin{aligned} f(X) &= D - 1/X \\ f'(X) &= 1/X^2 \end{aligned}$$

Plugging these value into the Newton-Raphson iterative equation gives

$$\begin{aligned} X_{i+1} &= X_i - \frac{f(X_i)}{f'(X_i)} \\ &= X_i - \left(\frac{D - 1/X_i}{1/X_i^2} \right) \\ &= X_i - D \cdot X_i^2 + X_i \\ &= 2 \cdot X_i - D \cdot X_i^2 \\ &= X_i \cdot (2 - D \cdot X_i) \end{aligned}$$

Each iteration requires one multiplication and one subtraction. Replacing the subtraction by a complement operation results in a small amount of addition error. An example of Newton-Raphson division is shown in Table 7.1, for $X = 1.875$, $D = 1.625$, $X_0 = 0.75$ where D is the divisor and X is the dividend and X_0 is the approximation. A graphical interpretation of this method for reciprocal is shown in Figure 7.1 where the derivative is utilized to iteratively find where the next value on the plot is found. Therefore, the final result is

i	X_i	$D \cdot X_i$	$2 - D \cdot X_i$
0	0.75	1.218750	0.78125
1	0.585938	0.952148	1.047852
2	0.613983	0.997726	1.002274
3	0.615372	0.999985	1.000015
4	0.615387		

Table 7.1. Newton-Raphson Division

computed by multiplying 0.615387 as shown in Table 7.1 by the value of the dividend. Since the algorithm computes the reciprocal, architectures can also encode the instruction set architecture to support reciprocal instructions.

$$Q = X \cdot X_4 = 1.875 \cdot 0.615387 = 1.153854$$

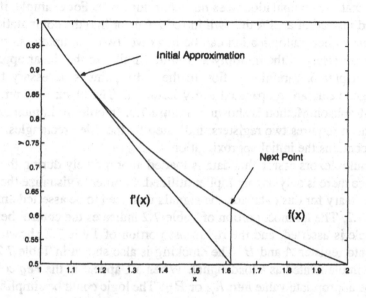

Figure 7.1. Newton-Raphson Iteration.

The absolute error in the approximation $X_i \approx 1/D$ is

$$\epsilon_{X_i} = X_i - 1/D$$

which gives

$$X_i = 1/D + \epsilon_{X_i}$$

Replacing X_i by $1/D + \epsilon_{X_i}$ in the Newton-Raphson division equation produces the following:

$$
\begin{aligned}
x_{i+1} &= X_i \cdot (2 - D \cdot X_i) \\
&= (1/D + \epsilon_{X_i}) \cdot (2 - D \cdot (1/D + \epsilon_{X_i})) \\
&= (1/D + \epsilon_{X_i}) \cdot (1 - D \cdot \epsilon_{X_i}) \\
&= 1/D - D \cdot \epsilon_{X_i}^2
\end{aligned}
$$

Therefore, the absolute error decreases quadratically as

$$\epsilon_{x_{i+1}} = -D \cdot \epsilon_{X_i}^2$$

If D is chosen such that $1 \leq D < 2$, and X_i is accurate to p fractional bits (i.e., $\mid \epsilon_{X_i} \mid < 2^{-p}$ then $\mid \epsilon_{X_{i+1}} \mid < 2^{-2 \cdot p}$. This means that each iteration approximately doubles the number of accurate bits in X_i as stated earlier. For example, if the accuracy of an initial approximation is 8 fractional bits, on the subsequent iteration the result will be accuracy will be 16 bits. However, this assumes that the computation does not incur any error. For example, if a CSA is utilized instead of a CPA this will introduce error into the computation.

Because either multiplication can be negative, two's complement multiplication is necessary in the multiply-add unit. Similar to the linear approximation in Chapter 6, careful attention to the radix point is necessary to make sure proper value are propagated every iteration. The block diagram for the Newton-Raphson Method is shown in Figure 7.2. In order to implement Newton Raphson requires two registers, indicated by the filled rectangles. The IA module contains the initial approximation.

The multiplexors assure that data is loaded appropriately during the correct cycle since there is only one multiplier utilized. In order to visualize the control logic necessary for this datapath, the signals that need to be asserted are shown in Table 7.2. The leftmost portion of Table 7.2 indicates the control before the clock cycle is asserted, and the rightmost portion of Table 7.2 shows what is written into register A and B. The clocking is also shown in Table 7.2 which could be implemented as a load signal. When a 1 appear in the reg column, it writes the appropriate value into R_A or R_B. The logic could be simplified with two multipliers. However, since multipliers consume a significant amount of memory, having two multipliers is normally not an option.

| | | | Before | | | | | After | |
| | | | mux | | | reg | | | |
Cycle	MCAN	MPLIER	A	B	D	A	B	R_A	R_B
1	X_0	D	1	0	1	1	1	$X_0 \cdot D$	$2 - X_0 \cdot D$
2	X_0	$2 - X_0 \cdot D$	1	1	1	1	0	X_1	$2 - X_0 \cdot D$
3	X_1	D	0	0	1	0	1	X_1	$2 - X_1 \cdot D$
4	X_1	$2 - X_1 \cdot D$	0	1	1	1	0	X_2	$2 - X_1 \cdot D$
5	X_2	D	0	0	1	0	1	X_2	$2 - X_2 \cdot D$
6	X_2	$2 - X_2 \cdot D$	0	1	1	1	0	X_3	$2 - X_2 \cdot D$
7	X_3	D	0	0	1	0	1	X_3	$2 - X_3 \cdot D$
8	X_3	$2 - X_3 \cdot D$	0	1	1	1	0	X_4	$2 - X_3 \cdot D$
9	X_4	X	0	0	0	1	0	$X_4 \cdot X$	$2 - X_3 \cdot D$

Table 7.2.   Control Logic for Datapath

In Figure 7.3, the Verilog code for the implementation is shown. For this implementation, a constant approximation is utilized, therefore, the initial ap-

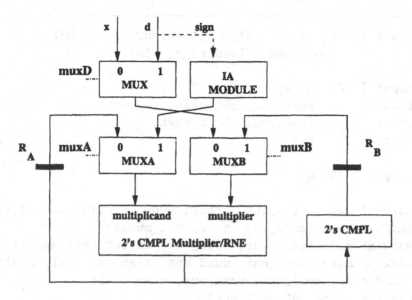

Figure 7.2.   Newton-Raphson Division Block Diagram.

proximation or X_0 (assuming $1 \leq d < 2$) is equal to $(1 + 1/2)/2 = 0.75$. The algorithm can handle two's complement numbers if the hardware decodes whether the divisor is negative. If the divisor is negative, the IA module should produce -0.75 as its approximation. A 2-1 multiplexor is utilized to choose the correct divisor in Figure 7.3 by using the sign bit of the divisor as the select signal for the multiplexor. The negative IA is chosen because the negative plot of $-1/x$ should have a negative approximation to start the iterative process. As mentioned previously, the instantiated multiplier in Figure 7.3 is a two's complement multiplier to handle both negative and positive numbers. The *twos_compl* module performs two's complementation because the Newton-Raphson equation requires $(2 - D \cdot X_i)$. In order to guarantee the correct result, the most significant bit of $D \cdot X_i$ is not complemented. In other words, simple two's complementation is implemented where the one's complement of $D \cdot X_i$ is taken and an ulp added. However, the most significant bit of $D \cdot X_i$ is not complemented. This guarantees the correct result (.e.g $0.75 = 00.1100 \rightarrow 01.0011 + ulp = 01.0100 = 1.25$). Since this implementation involves a simple row of $(m - 1)$ inverters where m is the internal precision within the unit and a CPA, the code is not shown. Similar to the recurrence dividers, the internal precision within the unit is typically larger than the input operand. Rounding logic (specifically RNE) is utilized to round the logic appropriately. Although RNE is performed by a separated module, it could easily be integrated within the multiplier [BK99].

```
module nrdiv (q, d, x, sel_mux, sel_muxa, sel_muxb,
              load_rega, load_regb, clk);

    input [7:0]  d, x;
    input        sel_muxa, sel_muxb, sel_mux;
    input        load_rega, load_regb;
    input        clk;

    output [7:0] q;

    mux21x8 mux_ia (ia_out, 8'b00110000, 8'b11010000, d[7]);
    mux21x8 muxd (muxd_out, x, d, sel_muxd);
    mux21x8 muxa (muxa_out, rega_out, ia_out, sel_muxa);
    mux21x8 muxb (muxb_out, muxd_out, regb_out, sel_muxb);
    csam8 csam0 (mul_out, muxa_out, muxb_out);
    rne rne0 (rne_out, mul_out);
    twos_compl tc0 (twoscmp_out, rne_out);
    register8 regb (regb_out, twoscmp_out, load_regb);
    register8 rega (rega_out, rne_out, load_rega);
    assign q = rne_out;

endmodule // nrdiv
```

Figure 7.3. Newton-Raphson Division Using Initial Approximation Verilog Code.

Since the IA module utilizes an initial approximation using a constant approximation, the precision of this approximation is only accurate to 2 bits. Since each iteration approximately doubles the number of bits of accuracy, only 2 to 3 iterations are necessary to guarantee a correct result (i.e. $2 \rightarrow 4 \rightarrow 8 \rightarrow 16$. In general, the number of iterations, p, where n is the desired precision and m is the accuracy of the estimate is:

$$p = \lceil log_2 \frac{n}{m} \rceil$$

7.2   Multiplicative-Divide Using Convergence

The Division by convergence iteration is used to improve division by utilizing what is called *iterative* convergence [Gol64]. For this method, the goal is to find a sequence, $K_1, K_2, \ldots$ such that the product $r_i = D \cdot K_1 \cdot K_2 \cdot \ldots K_i$ approaches 1 as i goes to infinity. That is,

$$q_i = X \cdot K_1 \cdot K_2 \cdot \ldots K_i \rightarrow Q$$

In other words, the iteration attempts to reduce the denominator to 1 while having the numerator approach X/D. To achieve this, the algorithm multiplies the top and bottom of the fraction by a value until the numerator converges to X/D.

As stated previously, the division by convergence division algorithm provides a high-speed method for performing division using multiplication and subtraction. The algorithm computes the quotient $Q = X/D$ using three steps [EIM+00]:

1 Obtain an initial reciprocal approximation, $K_1 \approx 1/D$.

2 Perform an iterative improvement of the new numerator (k times), $q_i = q_{i-1} \cdot K_i$, such that $q_0 = X$

3 Perform an iterative improvement of the denominator (k times), where $r_i = r_{i-1} \cdot K_i$), such that $r_0 = D$ **AND** Get ready for the next iteration, by normalizing the denominator by performing $K_{i+1} = (2 - r_i)$

The division by convergence algorithm is sometimes referred to Goldschmidt's division [Gol64]. An example of Goldschmidt's division is shown in Table 7.3, for $X = 1.875$, $D = 1.625$, $K_1 = 0.75$ where D is the divisor and X is the dividend and K_1 is the approximation. As opposed to the Newton-Raphson method, a final multiplication is not needed to complete the operation. Therefore, the quotient is $Q = 1.153839$.

i	q_i	r_i	$K_{i+1} = 2 - r_i$
0	1.406250	1.218750	0.781250
1	1.098633	0.952148	1.047852
2	1.151199	0.997711	1.002289
3	1.153839	1.000000	1.00000

Table 7.3. Goldschmidt's Division

The block diagram of the division by convergence is shown in Figure 7.4. Since division by convergence requires a multiplier, some designs incorporate the multiplier within the unit. For example, multiplication is performed in Figure 7.4 when $muxA = 1$ and $muxB = 1$. In Figure 7.5, the Verilog code for the implementation is shown. Similar to the Newton-Raphson implementation, the same constant approximation and the two's complement are utilized. The control logic, similar to the Newton-Raphson control, is shown in Table 7.4 Since the division by convergence does not need a final multiplication to produce the quotient as shown in Table 7.4, three registers are required to store the intermediate values.

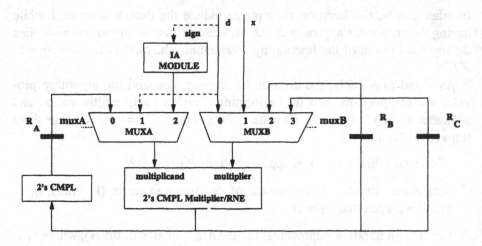

Figure 7.4. Goldschmidt's Division Block Diagram.

			Before					After		
			mux		reg					
Cycle	MCAN	MPLIER	A	B	A	B	C	R_A	R_B	R_C
1	K_1	X	2	1	0	1	0	0	$K_1 \cdot X$	0
2	K_1	D	2	0	1	0	1	K_2	$K_1 \cdot X$	$K_1 \cdot D$
3	K_2	Q_1	0	2	0	1	0	K_2	$K_2 \cdot Q_1$	$K_1 \cdot D$
4	K_2	R_1	0	3	1	0	1	K_3	$K_2 \cdot Q_1$	$K_2 \cdot R_1$
5	K_3	Q_2	0	2	0	1	0	K_3	$K_3 \cdot Q_2$	$K_2 \cdot R_1$
6	K_3	R_2	0	3	1	0	1	K_4	$K_3 \cdot Q_2$	$K_3 \cdot R_2$
7	K_4	Q_3	0	2	0	1	0	K_4	$K_4 \cdot Q_3$	$K_3 \cdot R_2$
8	K_4	R_3	0	3	1	0	1	K_5	$K_4 \cdot Q_3$	$K_4 \cdot R_2$

Table 7.4.   Control Logic for Datapath

7.3   Summary

The methods presented here provide an alternative to methods such as digit recurrence. The choice of implementation depends on many factors since many of the iterative approximation methods utilized a significant amount of hardware. Some implementations have suggested utilizing better approximations as well as better hardware for multiplying and accumulating [SSW97]. Iterative methods have also been successfully implemented within general-purpose processors [OFW99]. On the other hand, because multiplicative-divide algorithms converge quadratically IEEE rounding for the final quotient is more involved [Sch95].

```
module divconv (q, d, x, sel_muxa, sel_muxb,
                load_rega, load_regb, load_regc);

    input [7:0]  d, x;
    input [1:0]  sel_muxa, sel_muxb;
    input        load_rega, load_regb, load_regc;

    output [7:0] q;

    mux21x8 mux_ia (ia_out, 8'b00110000, 8'b11010000, d[7]);
    mux41x8 mux2 (muxb_out, d, x, regb_out, regc_out,
                  sel_muxb);
    mux31x8 mux3 (muxa_out, rega_out, d, ia_out, sel_muxa);
    csam8 csam0 (mul_out, muxa_out, muxb_out);
    rne rne0 (rne_out, mul_out);
    twos_compl tc0 (twoscmp_out, rne_out);
    register8 regc (rega_out, twoscmp_out, load_rega);
    register8 regb (regb_out, rne_out, load_regb);
    register8 rega (regc_out, rne_out, load_regc);
    assign q = rne_out;

endmodule // divconv
```

Figure 7.5.　Division by Convergence Using Initial Approximation Verilog Code.

References

[Atk68]   D. E. Atkins. Higher radix division using estimates of the divisor and partial remainder. *IEEE Transactions on Computer*, C-17:925–934, 1968.

[Avi61]   A. Avizienis. Signed-Digit number representations for fast parallel arithmetic. *IRE Transactions on Electronic Computers*, 10:389–400, 1961.

[Ber03]   J. Bergeron. *Writing Testbenches : Functional Verification of HDL Models, 2nd Edition*. Kluwer Academic Publishers, 2003.

[Bew94]   G. W. Bewick. *Fast multiplication: algorithms and implementation*. PhD thesis, Stanford University, 1994.

[BH01]   N. Burgess and C. Hinds. Design issues in radix-4 SRT square root and divide unit. In *Proceedings of the 35th Asilomar Conference on Signals, Systems, and Computers*, volume 2, pages 1646–1650, 2001.

[BK82a]   R. P. Brent and H. T. Kung. A regular layout for parallel adders. *IEEE Transaction on Computers*, C-31:260–264, 1982.

[BK82b]   R.P. Brent and H.Y. Kung. A regular layout for parallel adders. *IEEE Transactions on Computer*, C-31:260–264, 1982.

[BK99]   N. Burgess and S. Knowles. Efficient implementation of rounding units. In *Proceedings of the 33rd Asilomar Conference on Signals, Systems, and Computers*, volume 2, pages 1489–1493, 1999.

[Boo51]   A. D. Booth. A signed binary multiplication technique. *Q. J. Mech. Appl. Math.*, 4:236–240, 1951.

[BSL01]   A. Beaumont-Smith and C.-C Lim. Parallel-prefix adder design. In *Proceedings of the 15th Symposium on Computer Arithmetic*, pages 218–225, 2001.

[BSS95]   K. Bickerstaff, M. J. Schulte, and E. E. Swartzlander, Jr. Parallel reduced area multipliers. *Journal of VLSI Signal Processing*, 9:181–192, April 1995.

[BSS01]   K. Bickerstaff, M. J. Schulte, and E. E. Swartzlander, Jr. Analysis of column compression multipliers. In *Proceedings of the 15th Symposium on Computer Arithmetic*, pages 33–39, 2001.

[Bur02]   N. Burgess. The flagged prefix adder and its application in integer arithmetic. *Journal of VLSI Signal Processing*, 31(3):263–271, 2002.

[BW73]   C. R. Baugh and B. A. Wooley. A two's complement parallel array multiplication algorithm. *IEEE Transactions on Computers*, C-22:1045–1047, 1973.

[CGaLL+91]   B. W. Char, K. O. Geddes, G. H. Gonnet andB. L. Leong, M. B. Monagan, and S. M. Watt. *Maple V Library Reference Manual*. Springer Verlag, 1991.

[CSTO92]   P. K. Chan, M. D. Schlag, C. D. Thomborson, and V. G. Oklobdzija. Delay optimization of carry-skip adders and block carry-lookahead adders using multidimensional dynamic programming. *IEEE Transactions on Computers*, 41(8):920–930, 1992.

[CT95]   T. Coe and P.T.P. Tang. It takes six ones to reach a flaw. In *Proceedings of the 12th Symposium on Computer Arithmetic*, pages 140–146, 1995.

[CW80]   W. Cody and W. Waite. *Software Manual for the Elementary Functions*. Prentice-Hall, 1980.

[Dad65]   L. Dadda. Some schemes for parallel multipliers. *Alta Frequenza*, 34:349–356, 1965.

[EB99]   J. Eyre and J. Bier. DSP processors hit the mainstream. *IEEE Computer*, pages 51–59, 1999.

[EB00]   J. Eyre and J. Bier. The evolution of DSP processors. *IEEE Signal Processing Magazine*, pages 43–51, 2000.

[EIM+00]   M.D. Ercegovac, L. Imbert, D. Matula, J.-M. Muller, and G. Wei. Improving goldschmidt division, square root and square root reciprocal. *IEEE Transactions on Computers*, 49(7):759–763, 2000.

[EL90]   M. D. Ercegovac and T. Lang. Fast multiplication without carry-propagate addition. *IEEE Transactions on Computers*, 39(11):1385–1390, 1990.

[EL92a]   M. D. Ercegovac and T. Lang. Fast arithmetic for recursive computations. *VLSI Signal Processing V*, pages 14–18, 1992.

[EL92b]   M. D. Ercegovac and T. Lang. On-the-fly rounding. *IEEE Transactions on Computer*, C-41(12):1497–1503, 1992.

[EL94]   M. D. Ercegovac and T. Lang. *Division and Square Root: Digit-Recurrence Algorithms and Implementations*. Kluwer Academic Publishers, 1994.

[EL03]   M. D. Ercegovac and T. Lang. *Digital Arithmetic*. Morgan Kaufmann Publishers, 2003.

[FG00]   J. Fridman and Z. Greenfield. The TigerSHARC DSP architecture. *IEEE Micro*, 20(1):66–76, 2000.

[Fly70]   M. J. Flynn. On division by functional iteration. *IEEE Transactions on Computer*, C-19:702–706, 1970.

[GK83]   D. D. Gajski and R. H. Kuhn. New VLSI tools. *IEEE Computer*, pages 11–14, 1983.

[Gol64]   R. E. Goldschmidt. Application of division by convergence. Master's thesis, Massachusetts Institute of Technology, June 1964.

[Gol91]   D. Goldberg. What every computer scientist should know about floating-point arithmetic. *ACM Computing Surveys*, 23:5–48, 1991.

[GS03]   J. Grad and J. E. Stine. A standard cell library for student projects. In *Proceedings of the IEEE International Microelectronic System Education*, pages 98–99, 2003.

[GSH03]   J. Grad, J. E. Stine, and D. Harris. Hybrid EMODL Ling Addition. *Submitted to IEEE Transactions in Solid State Circuits*, 2003.

[GSss]   J. Grad and J. E. Stine. Hybrid EMODL Ling addition. In *Proceedings of the 36th Asilomar Conference on Signals, Systems, and Computers*, 2002 (in press).

[HC87]   T. Han and D. A. Carlson. Fast area-efficient VLSI adders. In *Proceedings of the 8th Symposium on Computer Arithmetic*, pages 49–56, 1987.

[Hig94]   N. J. Higham. *Accuracy and Stability of Numerical Algorithms*. SIAM, 1994.

[HT95]   H. Hassler and N. Takagi. Function evaluation by table look-up and addition. In *Proceedings of the 12th Symposium on Computer Arithmetic*, pages 10–16, 1995.

[HW70]   A. Habibi and P. A. Wintz. Fast multipliers. *IEEE Transactions on Computers*, C-19:153–157, 1970.

[IEE85]   IEEE Computer Society. *IEEE Standard 754 for Binary Floating Point Arithmetic*. IEEE Press, August 1985.

[IEE95]   IEEE Computer Society. *IEEE Standard Hardware Descriptive Language Based on the Verilog Hardware Descriptive Language*. IEEE Press, 1995.

[IEE01]   IEEE Computer Society. *IEEE Standard Verilog Hardware Description Language*. IEEE Press, 2001.

[IEE02]   IEEE Computer Society. *IEEE Standard for Verilog Register Transfer Level Synthesis*. IEEE Press, 2002.

[KA97]   P. Kurup and T. Abbasi. *Logic Synthesis with Synopsys*. Kluwer Academic Press, 1997.

[Kno99]   S. Knowles. A family of adders. In *Proceedings of the 14th Symposium on Computer Arithmetic*, pages 30–34, 1999.

[Kno01]   S. Knowles. A family of adders. In *Proceedings of the 15th Symposium on Computer Arithmetic*, pages 277–281, 2001.

[Kor93]   I. Koren. *Computer Arithmetic and Algorithms*. Prentice Hall, 1993.

[KS73]   P. M. Kogge and H. S. Stone. A parallel algorithm for the efficient solution of a general class of recurrence equations. *IEEE Transactions on Computers*, C-22:783–791, 1973.

[KS98]   E. J. King and E. E. Swartzlander, Jr. Data-dependent truncated scheme for parallel multiplication. In *Proceedings of the Thirty First Asilomar Conference on Signals, Circuits and Systems*, pages 1178–1182, 1998.

[LA94]   H. Lindkvist and P. Andersson. Techniques for fast CMOS-based conditional sum adders. In *Proceedings of the 1994 International Conference on Computer Design*, pages 626–635, October 1994.

[LF80]   R. E. Ladner and M. J. Fischer. Parallel prefix computation. *Journal of the ACM*, 27(4):831–838, October 1980.

[Lim92]   Y.C. Lim. Single precision multiplier with reduced circuit complexity for signal processing applications. *IEEE Transactions on Computers*, 41(10):1333–1336, 1992.

[LMT98]   V. Lefèvre, J.-M. Muller, and A. Tisserand. Toward correctly rounded transcendentals. *IEEE Transactions on Computers*, 47(11):1235–1243, 1998.

[LS92]   T. Lynch and E. E. Swartzlander, Jr. A spanning tree carry lookahead adder. *IEEE Transactions on Computer*, C-41(8):931–939, 1992.

[Mac61]   O. L. MacSorley. High-speed arithmetic in binary computers. *IRE Proceedings*, 49:67–91, 1961.

[Mat87]   J. H. Mathews. *Numerical Methods for Computer Science, Engineering and Mathematics*. Prentice Hall, 1987.

[MK71]   J. C. Majithia and R. Kitai. An iterative array for multiplication of signed binary numbers. *IEEE Transactions on Computers*, C-20(2):214–216, February 1971.

[MT90]   G.-K. Ma and F. J. Taylor. Multiplier policies for digital signal processing. *IEEE ASSP Magazine*, 7(1):6–19, 1990.

[Mul97]   J.-M. Muller. *Elementary Function, Algorithms and Implementation*. Birkhauser Boston, 1997.

[NL99]   A. Nannarelli and T. Lang. Low-power division: comparison among implementations of radix 4, 8 and 16. In *Proceedings of the 14th Symposium on Computer Arithmetic*, pages 60–67, 1999.

[oCB92]   University of California-Berkeley. Berkeley Logic Interchange Format (BLIF). Technical report, University of California-Berkeley, 1992.

[OF97]   S. F. Oberman and M. J. Flynn. Design issues in division and other floating-point operations. *IEEE Transactions on Computers*, 46(2):154–161, 1997.

[OF98]   S. F. Oberman and M. J. Flynn. Minimizing the complexity of SRT tables. *IEEE Transactions on Very Large Scale Integration Systems*, 6(1):141–149, 1998.

[OFW99]   S. Oberman, G. Favor, and F. Weber. AMD 3DNow! technology: architecture and implementations. *IEEE Micro*, 19:37–48, 1999.

[Par90]   B. Parhami. Generalized Signed-Digit number systems: A unifying framework for redundant number representations. *IEEE Transactions on Computers*, C-39(1):89–98, January 1990.

[Par01]   K. K. Parhi. Approaches to low-power implementations of DSP systems. *IEEE Transactions on Circuits and Systems I: Fundamental Theory and Applications*, 48(10):1214–1224, 2001.

[Pez71]   S. D. Pezaris. A 40-ns 17-bit by 17-bit array multiplier. *IEEE Transactions on Computers*, C-20:442–447, 1971.

[PZ95]   A. Prabhu and G. Zyner. 167 mhz radix-8 divide and square root using overlapped radix-2 stages. In *Proceedings of the 12th Symposium on Computer Arithmetic*, pages 155–162, 1995.

[Rob58]   J. E. Robertson. A new class of digital division methods. *IRE Transactions on Electronic Computers*, EC-7:218–222, 1958.

[Sch95]   E. Schwarz. Rounding for quadratically converging algorithms for division and square root. In *Proceedings of the 29th Asilomar Conference on Signals, Systems, and Computers*, volume 1, pages 600–603, 1995.

[Sch03]   E. Schwarz. Revisions to the IEEE 754 standard for floating-point arithmetic. In *Proceedings of the 16th Symposium on Computer Arithmetic*, pages 112–112, 2003.

[SD03]   J. E. Stine and O. M. Duverne. Variations on truncated multiplication. In *Euromicro Symposium on Digital System Design*, pages 112–119, 2003.

[Ses98]   N. Seshan. High VelociTI processing. *IEEE Signal Processing*, 15(2):86–101, 1998.

[SL95]   P. Soderquist and M. Leeser. An area/performance comparison of subtractive and multiplicative divide/square root implementations. In *Proceedings of the 12th Symposium on Computer Arithmetic*, pages 132–139, 1995.

[SM93]   D. D. Sarma and D. W. Matula. "Measuring the accuracy of ROM reciprocal tables. In *Proceedings of the 11th Symposium on Computer Arithmetic*, pages 95–102, July 1993.

[SM95]   D. D. Sarma and D. W. Matula. Faithful bipartite ROM reciprocal tables. In *Proceedings of the 12th Symposium on Computer Arithmetic*, pages 17–29, 1995.

[SP92]   H. R. Srinivas and K. K. Parhi. A fast VLSI adder architecture. *IEEE Journal of Solid-State Circuits*, 27(5):761–767, May 1992.

[SP94]   H. R. Srinivas and K. K. Parhi. A fast radix 4 division algorithm. In *IEEE International Symposium on Circuits and Systems*, pages 311–314, 1994.

[SS93]   M. J. Schulte and E. E. Swartzlander, Jr. Truncated multiplication with correction constant. In *VLSI Signal Processing VI*, pages 388–396, October 1993.

[SS98]   J. E. Stine and M. J. Schulte. The symmetric table addition method for accurate function approximation. *Journal of VLSI Signal Processing*, 21(2):167–177, 1998.

[SS99]   M. J. Schulte and J. E. Stine. Approximate elementary functions with symmetric bipartite tables. *IEEE Transactions on Computers*, 48(8):842–847, 1999.

[SSH99]   I. Sutherland, R. F. Sproull, and D. Harris. *Logical Effort : Designing Fast CMOS circuits*. Morgan Kaufmann Publishers, 1999.

[SSL+92]   E. M. Sentovich, K. J. Singh, L. Lavagno, C. Moon, R. Murgai, A. Saldanha, H. Savoj, P. R. Stephan, R. K. Brayton, and A. L Sangiovanni-Vincentelli. SIS: A system for sequential circuit synthesis). Technical Report UCB/ERL M92/41, University of California-Berkeley, 1992.

[SSW97]   M. J. Schulte, J. E. Stine, and K. E. Wires. High-speed reciprocal approximations. In *Proceedings of the 31st Asilomar Conference on Signals, Systems, and Computers*, 1997.

[Sut01]   S. Sutherland. *Verilog-2001 A Guide to the New Features of the Verilog Hardware Description Language*. Kluwer Academic Publishers, 2001.

[Swa80]   E. E. Swartzlander, Jr. Merged arithmetic. *IEEE Transactions on Computers*, C-29:946–950, 1980.

[Swa90a]   E. E. Swartzlander, Jr. *Computer Arithmetic I*. IEEE Press, 1990.

[Swa90b]   E. E. Swartzlander, Jr. *Computer Arithmetic II*. IEEE Press, 1990.

[Tak92]   N. Takagi. A radix-4 modular multiplication hardware algorithm for modular exponentiation. *IEEE Transactions on Computers*, C-41(8):949–956, 1992.

[Tay85]   G. Taylor. Radix 16 srt dividers with overlapped quotient selection stages. In *Proceedings of the 7th Symposium on Computer Arithmetic*, pages 64–71, 1985.

[Toc58]   K. D. Tocher. Techniques of multiplication and division for automatic binary computers. *Quarterly Journal of Mechanics and Applied Mathematics*, 11:364–384, 1958.

[Vol59]   Jack E. Volder. The CORDIC trigonometric computing technique. *IRE Transactions on Electronic Computers*, EC-8:330–334, 1959.

[Wal64]   C. S. Wallace. Suggestion for a fast multiplier. *IEEE Transactions on Electronic Computers*, EC-13:14–17, 1964.

[Wal71]   J. S. Walther. A Unified approach for elementary functions. In *Spring Joint Computer Conference*, pages 379–385, 1971.

[WE85]   N. Weste and K. Eshraghian. *Principles of CMOS VLSI Design*. Addison-Wesley, 1985.

[Wei82]   A. Weinberger. A 4:2 carry-save adder module. *IBM Technical Disclosure Bulletin*, 23(8):3811–3814, 1982.

[Win65]   S. Winograd. On the time required to perform addition. *Journal of the ACM*, 12(2):277–285, April 1965.

[Win68]   S. Winograd. How f1st can computers add? *Scientific American*, pages 93–100, October 1968.

[WJM+97]   Z. Wang, G. A. Jullien, W. C. Miller, J. Wang, and S. S. Bizzan. Fast adders using enhanced multiple-output domino logic. *IEEE Journal of Solid-State Circuits*, 32(2):206–214, 1997.

[WSS01]   K. E. Wires, M. J. Schulte, and J. E. Stine. Combined IEEE Compliant and Truncated Floating Point Multipliers for Reduced Power Dissipation. In *Proceedings of the International Conference on Computer Design*, pages 497–500, 2001.

[YLCL92]   S-M. Yen, C-S. Laih, C-H. Chen, and J-Y. Lee. An efficient redundant-binary number to binary number converter. *IEEE Journal of Solid-State Circuits*, 27(1):109–112, January 1992.

[Zim97]   R. Zimmermann. *Binary adder architectures for cell-based VLSI and their synthesis*. PhD thesis, Swiss Federal Institute of Technology Zurich, 1997. Available at http://www.iis.ee.ethz.ch/~ zimmi,.

[ZPK00]   B. Ziegler, H. Praehofer, and T. G. Kim. *Theory of Modeling and Simulation*. Academic Press, 2000.

Index

CD-ROM Disclaimer